Improved Estimates of Filtered Total Mercury Loadings and Total Mercury Concentrations of Solids from Potential Sources to Sinclair Inlet, Kitsap County, Washington

By Anthony J. Paulson, Kathleen E. Conn, and John F. DeWild

Prepared in cooperation with Department of the Navy
Naval Facilities Engineering Command, Northwest

Scientific Investigations Report 2013–5081

U.S. Department of the Interior
U.S. Geological Survey

U.S. Department of the Interior
SALLY JEWELL, Secretary

U.S. Geological Survey
Suzette M. Kimball, Acting Director

U.S. Geological Survey, Reston, Virginia: 2013

Contents

Figures

Conversion Factors, Datums, and Abbreviations and Acronyms

Conversions Factors

Inch/Pound to SI

Multiply	By	To obtain
Length		
inch (in.)	2.54	centimeter (cm)
foot (ft)	0.3048	meter (m)

SI to Inch/Pound

Multiply	By	To obtain
Length		
micrometer (μm)	0.003937	inch (in.)
millimeter (mm)	0.03937	inch (in.)
centimeter (cm)	0.3937	inch (in.)
meter (m)	3.281	foot (ft)
meter (m)	1.094	yard (yd)
Area		
square kilometer (km^2)	247.1	acre
square meter (m^2)	10.76	square foot (ft^2)
square kilometer (km^2)	0.3861	square mile (mi^2)
Volume		
milliliter (mL)	0.0338	ounce, fluid (fl. oz)
liter (L)	0.2642	gallon (gal)
cubic meter (m^3)	0.0002642	million gallons (Mgal)
cubic meter (m^3)	0.0008107	acre-foot (acre-ft)
Flow rate		
cubic meter per second (m^3/s)	70.07	acre-foot per day (acre-ft/d)
milliliter per minute (mL/min)	5.8857	cubic foot per second (ft^3/s)
Mass		
gram (g)	0.03527	ounce, avoirdupois (oz)
kilogram (kg)	2.205	pound, avoirdupois (lb)
metric ton per year	1.102	ton per year (ton/yr)
Sedimentation rate		
gram per square centimeter (g cm^{-2})	0.228	ounce per square inch (oz/in^2)

Temperature in degrees Celsius (°C) may be converted to degrees Fahrenheit (°F) as follows:

$$°F=(1.8×°C)+32$$

Temperature in degrees Fahrenheit (°F) may be converted to degrees Celsius (°C) as follows:

$$°C=(°F-32)/1.8$$

Specific conductance is given in microsiemens per centimeter at 25 degrees Celsius (μS/cm at 25 °C).

Concentrations of chemical constituents in water are given either in milligrams per liter (mg/L), micrograms per liter (μg/L), or nanograms per liter (ng/L).

Concentrations of chemical constituents of solids are given in either percentage of dry weight, milligrams per kilogram (mg/kg) or nanograms per milligram (ng/mg), which are equivalent.

Conversion Factors, Datums, and Abbreviations and Acronyms—Continued

Datums

Vertical coordinate information is referenced to the North American Vertical Datum of 29 (NAVD 29). The vertical datum mean lower low water (MLLW) is defined as -1.83 m relative to NAVD 29.

Horizontal coordinate information is referenced to the North American Datum of 1988 (NAD 88).

Altitude, as used in this report, refers to distance above the vertical datum.

Abbreviations and Acronyms

BNC	Bremerton naval complex
CERCLA	Comprehensive Environmental Response, Compensation, and Liability Act
ENVVEST	ENVironmental inVESTment
FTHg	Filtered total mercury (ng/L)
HCl	Hydrochloric acid
HDPE	High-density polyethylene
LTMP	Long-term Monitoring Program
MHg	Methylmercury (ng/L)
MLLW	Mean lower low water
NBK Bremerton	Naval Base Kitsap at Bremerton
OU	Operable unit
OUBT	Operable Unit B Terrestrial
PETG	Polyethylene terephthalate copolyester
PFA	Perfluoroalkoxy
PP	Polypropylene
PSNS	Puget Sound Naval Shipyard
PSNS&IMF	Puget Sound Naval Shipyard and Intermediate Maintenance Facility
PSNSXXX	Stormwater drain of Puget Sound Naval Shipyard (XXX = stormwater drain number)
PST	Pacific Standard Time
PTFE	Polytetrafluoroethylene
PTHg	particulate total mercury (ng/L)
QFF	Quartz fiber filter
RI/FS	Remedial Investigation/Feasibility Study
RO	Reverse-osmosis
RPD	Relative percent difference
SSC	Suspended sediment concentration (mg/L)
STHg	Total mercury concentration of sediment (mg/kg)
THg	Total mercury (ng/L)
TSS	Total suspended solids (mg/L)
USGS	U.S. Geological Survey
WAWSC	Washington Water Science Center
WMRL	Wisconsin Mercury Research Laboratory
WTHg	Total mercury in whole (unfiltered) water (ng/L)
YSI™	Yellow Springs Instruments

Improved Estimates of Filtered Total Mercury Loadings and Total Mercury Concentrations of Solids from Potential Sources to Sinclair Inlet, Kitsap County, Washington

By Anthony J. Paulson, Kathleen E. Conn, and John F. DeWild

Abstract

Previous investigations examined sources and sinks of mercury to Sinclair Inlet based on historic and new data. This included an evaluation of mercury concentrations from various sources and mercury loadings from industrial discharges and groundwater flowing from the Bremerton naval complex to Sinclair Inlet. This report provides new data from four potential sources of mercury to Sinclair Inlet: (1) filtered and particulate total mercury concentrations of creek water during the wet season, (2) filtered and particulate total mercury releases from the Navy steam plant following changes in the water softening process and discharge operations, (3) release of mercury from soils to groundwater in two landfill areas at the Bremerton naval complex, and (4) total mercury concentrations of solids in dry dock sumps that were not affected by bias from sequential sampling.

The previous estimate of the loading of filtered total mercury from Sinclair Inlet creeks was based solely on dry season samples. Concentrations of filtered total mercury in creek samples collected during wet weather were significantly higher than dry weather concentrations, which increased the estimated loading of filtered total mercury from creek basins from 27.1 to 78.1 grams per year.

Changes in the concentrations and loading of filtered and particulate total mercury in the effluent of the steam plant were investigated after the water softening process was changed from ion-exchange to reverse osmosis and the discharge of stack blow-down wash began to be diverted to the municipal water-treatment plant. These changes reduced the concentrations of filtered and particulate total mercury from the steam plant of the Bremerton naval complex, which resulted in reduced loadings of filtered total mercury from 5.9 to 0.15 grams per year.

Previous investigations identified three fill areas on the Bremerton naval complex, of which the western fill area is thought to be the largest source of mercury on the base. Studies of groundwater in the other two fill areas were conducted under worst-case higher high tidal conditions. A December 2011 study found that concentrations of filtered total mercury in the well in the fill area on the eastern boundary of the Bremerton naval complex were less than or equal to 11 nanograms per liter, indicating that releases from the eastern area were unlikely. In addition, concentrations of total mercury of solids were low (<3 milligrams per kilogram). In contrast, data from the November 2011 study indicated that the concentrations of filtered total mercury in the well located in the central fill area had tidally influenced concentrations of up to 500 nanograms per liter and elevated concentrations of total mercury of solids (29–41 milligrams per kilogram). This suggests that releases from this area, which has not been previously studied in detail, may be substantial.

Previous measurements of total mercury of suspended solids in the dry dock discharges revealed high concentration of total mercury when suspended-solids concentrations were low. However, this result could have been owing to bias from sequential sampling during changing suspended-solids concentrations. Sampling of two dry dock systems on the complex in a manner that precluded this bias confirmed that suspended-solids concentrations and total mercury concentrations of suspended solids varied considerably during pumping cycles. These new data result in revised estimates of solids loadings from the dry docks. Although most of the solids discharged by the dry docks seem to be recycled Operable Unit B Marine sediment, a total of about 3.2 metric tons of solids per year containing high concentrations of total mercury were estimated to be discharged by the two dry dock systems. A simple calculation, in which solids (from dry docks, the steam plant, and tidal flushing of the largest stormwater drain) are widely dispersed throughout Operable Unit B Marine, suggests that Bremerton naval complex solids would likely have little effect on Operable Unit B Marine sediments because of high concentrations of mercury already present in the sediment.

Introduction

As early as the 1980s, the sediment of Sinclair Inlet in Puget Sound, Washington (fig. 1), was known to have elevated concentrations of a number of contaminants (Malins and others, 1982). A remedial investigation of the marine waters off the Bremerton naval complex (BNC), Bremerton, Washington, was completed under the Comprehensive Environmental Response, Compensation, and Liability Act (CERCLA) in 1996 (U.S. Navy, 2002), and the Record of Decision (U.S. Environmental Protection Agency, 2000) was issued as final in 2002. The remediation options included isolating a considerable volume of contaminated sediment from interactions with the benthic food web by capping and disposing of dredge spoils in a covered confined aquatic disposal pit in 2001. The primary objective of the marine sediment cleanup was to address the potential risk to humans, particularly those engaged in a subsistence lifestyle, from consumption of bottom-dwelling fish with tissue containing polychlorinated biphenyls (U.S. Navy, 2002). Three pathways were identified as having the capability to transport chemicals from the terrestrial landscape of the BNC to the marine environment, and thus having the potential to recontaminate the recently remediated marine sediment. The pathways included direct dry dock discharges, discharge of groundwater, and discharge from stormwater drain systems handling surface-water runoff.

As lead agency for environmental cleanup of the BNC, the U.S. Navy has completed the second 5-year review of the remedial actions of the marine sediment within the boundary of the BNC (U.S. Navy, 2008a), pursuant to Section 121(c) of CERCLA and the National Oil and Hazardous Substances Pollution Contingency Plan (40 Code of Federal Regulations Part 300). One of the issues highlighted in the second 5-year review was:

> There is insufficient information to determine whether the remedial action taken at OU [Operable Unit] B Marine with respect to mercury in sediment is protective of ingestion of rockfish by subsistence finfishers, (U.S. Navy, 2008a, p. 5).

Recommendations and follow-up actions in the second 5-year review are as follows:

- Revisit Remedial Investigation/Feasibility Study (RI/FS) groundwater-to-surface-water transport evaluations in light of total mercury (THg) concentrations in two long-term monitoring wells.

- Perform trend analyses and assess functionality and protectiveness of remedy for marine sediment.

- Collect additional information necessary to perform a risk evaluation and reach conclusions regarding the protectiveness of the remedy (U.S. Navy, 2002) with respect to total mercury concentrations in Sinclair Inlet sediment and fish tissue.

Since 2007, the U.S. Geological Survey (USGS) and the U.S. Navy have entered into several multi-year interagency agreements. The objectives of these studies are to (1) estimate the magnitudes of the different predominant sources of THg to Sinclair Inlet, including those from the BNC, (2) evaluate the transformation of mercury to a bioavailable form in Sinclair Inlet, and (3) assess the effect of the sources and transformation processes on the mercury burden in marine organisms and sediment.

Definitions.—In this report, total mercury (THg, in ng/L) refers to all chemical forms of mercury (including methylmercury, MHg, in ng/L) and not to an unfiltered water sample, therefore, THg = MHg + inorganic mercury. The medium of the sample associated with the THg is identified as whole water (unfiltered) total mercury (WTHg, in ng/L), filtered total mercury (FTHg, in ng/L), and particulate total mercury (PTHg, in ng/L), where WTHg = FTHg + PTHg.

Total mercury concentrations of marine sediment (STHg, in mg/kg) were measured by the Long-Term Monitoring Program (LTMP) of U.S. Navy. The total suspended-solids (TSS) concentration is a measure of solids concentrations in a sample collected independently of the PTHg sample. Suspended-solids concentrations (SSC) is a measure of the solids concentration of a whole sample collected from a churn splitter in concert with collection of the PTHg sample. The THg concentration of suspended solids (ng/mg) is calculated as: PTHg (ng/L)/TSS (mg/L) or PTHg (ng/L)/SSC (mg/L). By conversion, the concentration units of THg of suspended solids (ng/mg) are equal to the concentration units of STHg (mg/kg).

Defining the hydraulic gradient between groundwater elevation measured on the terrestrial datum NAVD 29 and the marine surface of Sinclair Inlet measured on the marine datum mean lower low water (MLLW) requires conversion to a common vertical datum. In Sinclair Inlet, the zero elevation of the NAVD 29 datum is +1.83 m relative to MLLW. The terrestrial elevations (for example, water level of groundwater in monitoring wells or water in stormwater vaults) were converted to the marine MLLW datum by adding 1.83 m to the elevation in terrestrial NAVD 29 datum.

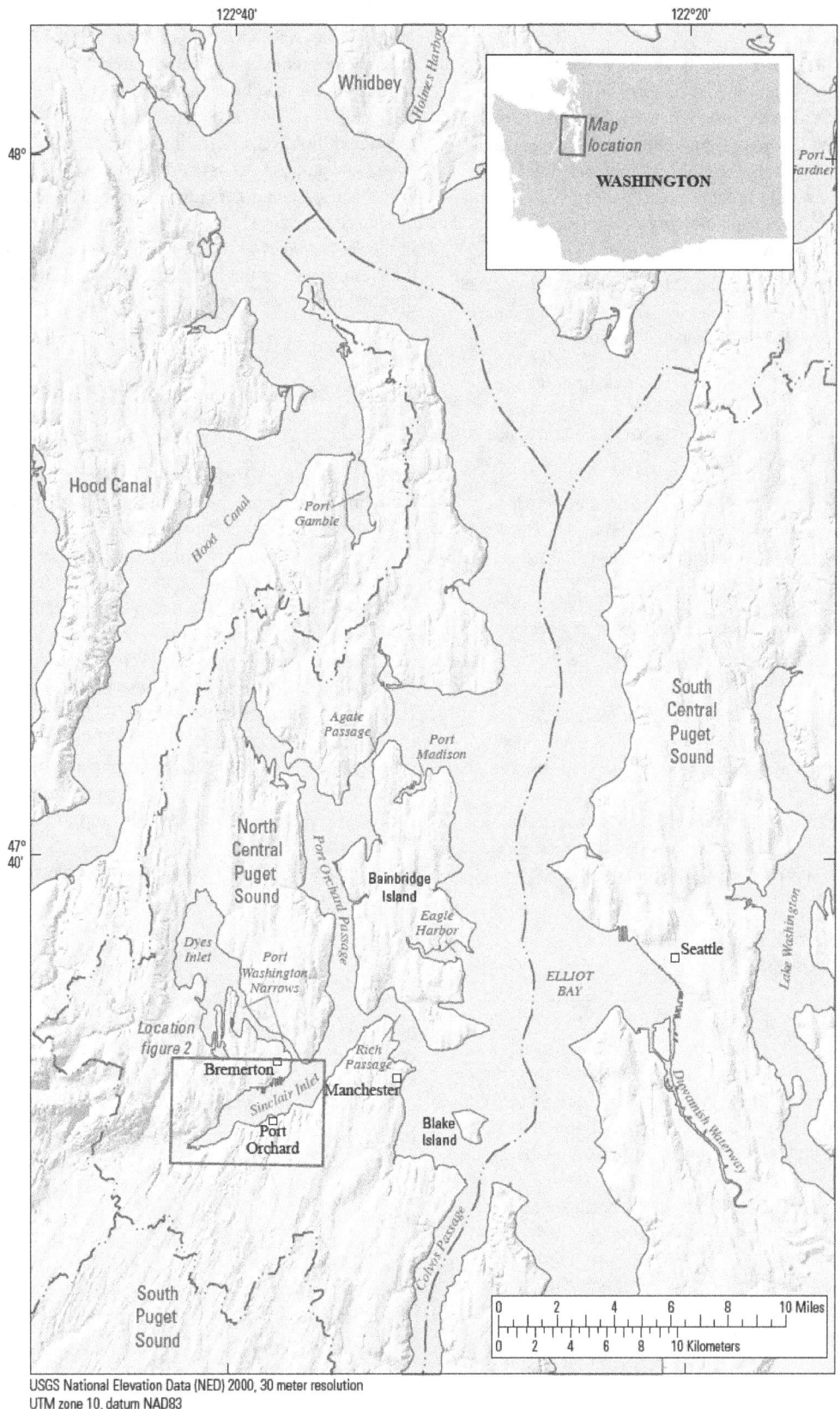

Figure 1. Location of Sinclair Inlet in Puget Sound, Kitsap County, Washington.

Purpose and Scope

As part of the Sources and Sinks study (Paulson and others, 2012), USGS examined the sources of THg to Sinclair Inlet and collected new data to evaluate mercury concentrations and loadings from industrial discharges and groundwater flowing from the BNC into Sinclair Inlet. This report provides new data to improve estimates of filtered total mercury loadings and/or total mercury concentrations of solids from four potential sources to Sinclair Inlet:

1. Measurements of FTHg and THg concentrations in creek samples collected during the wet season to complement measurements during dry season base flow in 2008.

2. Measurements of FTHg and THg concentrations in the steam plant discharge following changes in water softening process and discharge operations since the initial sampling in 2008.

3. Measurements of FTHg and THg concentrations in groundwater from two landfill areas on the BNC identified as Site 10C (central) and Site 10E (eastern) in the Preliminary Assessment (U.S. Navy, 1992).

4. Measurements of THg concentrations in dry docks solids that are not affected by sequential sampling during changing solids conditions.

These new data provide a recent and more accurate estimate of the sources of FTHg presented as part of the conceptual model of mercury behavior in Sinclair Inlet in Paulson and others (2012). In addition, new unbiased THg data of dry dock solids allow the first assessment of the net effect of discharges of solids from BNC sources on the concentrations of marine sediment within the BNC.

Site Description

Sinclair Inlet is a shallow (maximum depth of 20 m) embayment of Puget Sound on the west side of the Puget Sound lowland (fig. 1). Sinclair Inlet is defined in this study as the marine water body landward of the cable area that extends from the Bremerton ferry dock of the Washington State Ferry System on the north (fig. 2) to the mouth of Annapolis Creek on the south (fig. 2).

The BNC covers about 2 km² on the north shore of Sinclair Inlet in Bremerton, Washington (fig. 2) and includes the Puget Sound Naval Shipyard and Intermediate Maintenance Facility, Bremerton site (PSNS&IMF), and

Naval Base Kitsap at Bremerton (NBK Bremerton). The primary role of PSNS&IMF (1.5 km²) is to provide overhaul, maintenance, conversion, refueling, defueling, and repair services to the naval fleet and is a fenced, high-security area. The primary role of NBK Bremerton (0.4 km²) is to serve as a deep-draft home port for aircraft carriers and supply ships. For purposes of calculating fluxes of FTHg to Sinclair Inlet, Paulson and others (2012) divided the BNC into three discharge zones (fig. 3A): (1) the Zone of Direct Discharge in the western part of the BNC, (2) the Vicinity of Site 2 in the western part of the BNC, and (3) the Capture Zone of the Sumps of the dry docks.

Summary of Previous Results

Creeks.—From 2002–05, the ENVironmental inVESTment (ENVVEST) project sampled several creeks, including the four creeks within Sinclair Inlet Basin (Blackjack, Anderson, Annapolis and Gorst Creeks in fig. 2). Samples for wet season (March), dry season (September), and a range of storm conditions were collected (ENVVEST, 2006). WTHg concentrations ranged between 1.06 and 27.31 ng/L and generally correlated with TSS concentration (2–153 mg/L) which generally increased with wet weather. During USGS sampling of dry base flow conditions in 2008, the median concentrations of FTHg concentrations (n = 8) in the Sinclair Inlet creeks was 0.55 ng/L and ranged between 0.39 and 0.81 ng/L (Huffman and others, 2012). Median THg concentrations of suspended solids were 0.12 mg/kg (0.09–0.28 mg/kg). This report addresses wet weather conditions.

Steam Plant.—The steam plant is located on BNC in the Zone of Direct Discharge and discharges at the end of Mooring F into cell 26 (fig. 3A). In 2008, the effluent from the steam plant included backwash from the ion exchange columns of the demineralizing process and solutions from the stack blow-down process. FTHg concentrations of the steam plant effluent ranged between 15.3 and 143 ng/L and were correlated with specific conductance. TSS concentrations in the steam plant effluent were low (<1.5 mg/L) and the THg concentrations of solids ranged from 2.95 to 68.7 mg/kg, with the highest value corresponding to the lowest TSS value of 0.47 mg/L. After 2008, the steam plant converted to a reverse osmosis (RO) water softening process and the blow-down solutions were no longer discharged as part of the effluent but instead were sent to the wastewater treatment plant. This report addresses the changes in THg loads from the steam plant to Sinclair Inlet owing to the process changes.

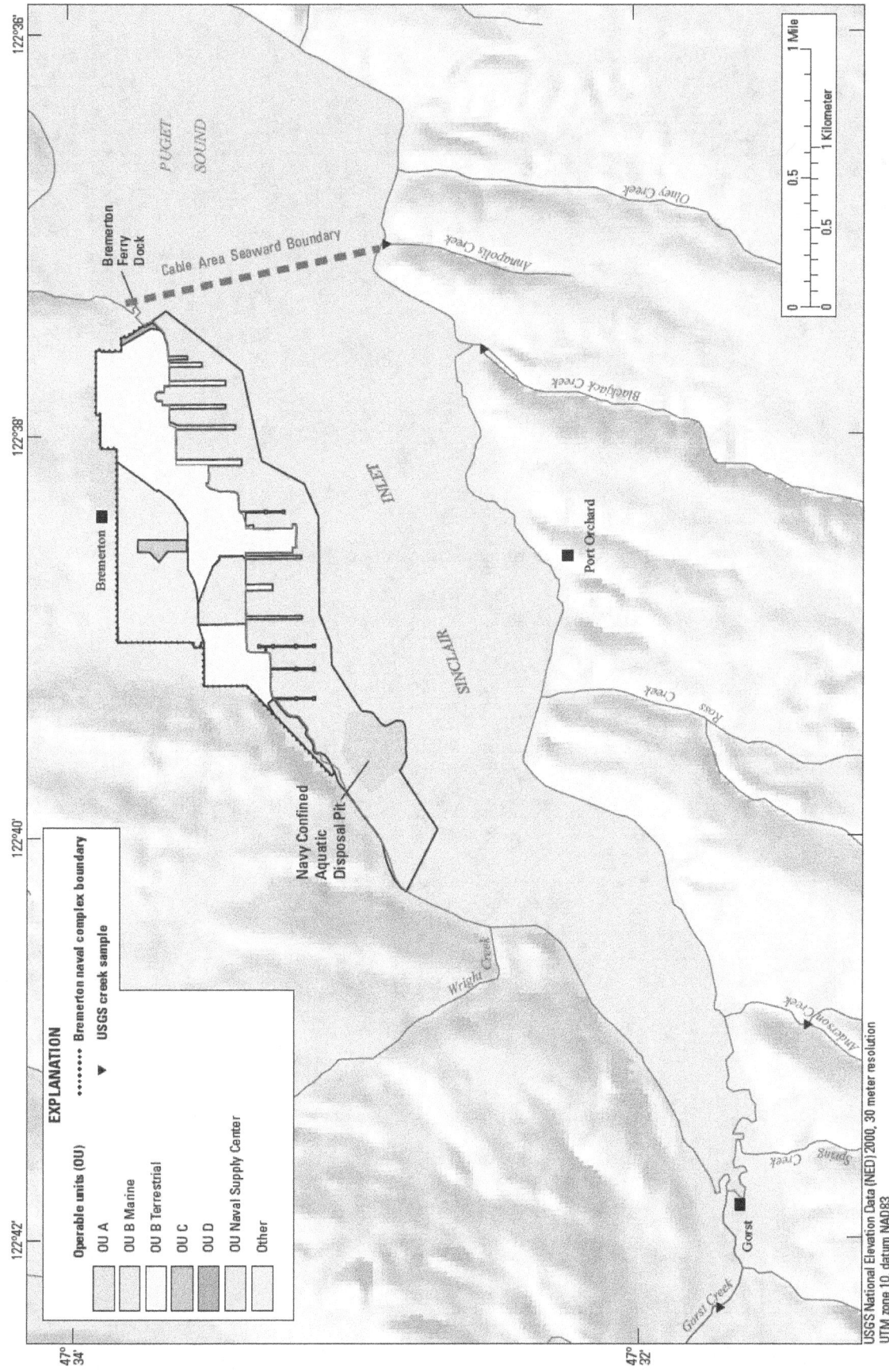

USGS National Elevation Data (NED) 2000, 30 meter resolution
UTM zone 10, datum NAD83

Figure 2. Locations of the Sinclair Inlet seaward boundary, U.S. Geological Survey creek sampling sites, and the Bremerton naval complex Operable Units and boundary, Sinclair Inlet, Kitsap County, Washington.

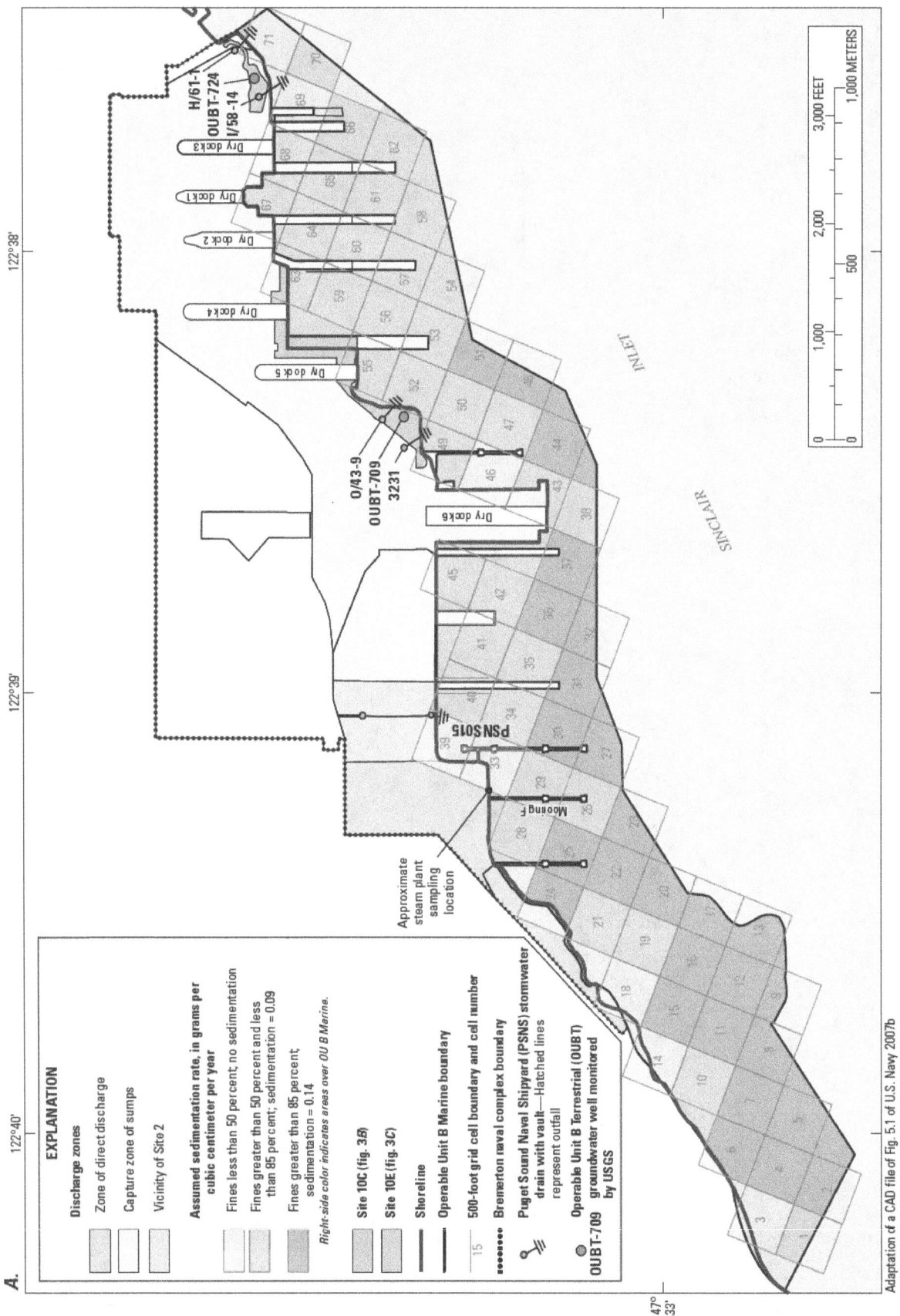

Figure 3. Location of (*A*) the U.S. Navy Long-Term Monitoring Program grid cells in Operable Unit (OU) B Marine and locations of U.S. Geological Survey (USGS) sampling sites within the Bremerton naval complex boundary, including detailed views of (*B*) Site 10C and (*C*) Site 10E, Sinclair Inlet, Kitsap County, Washington.

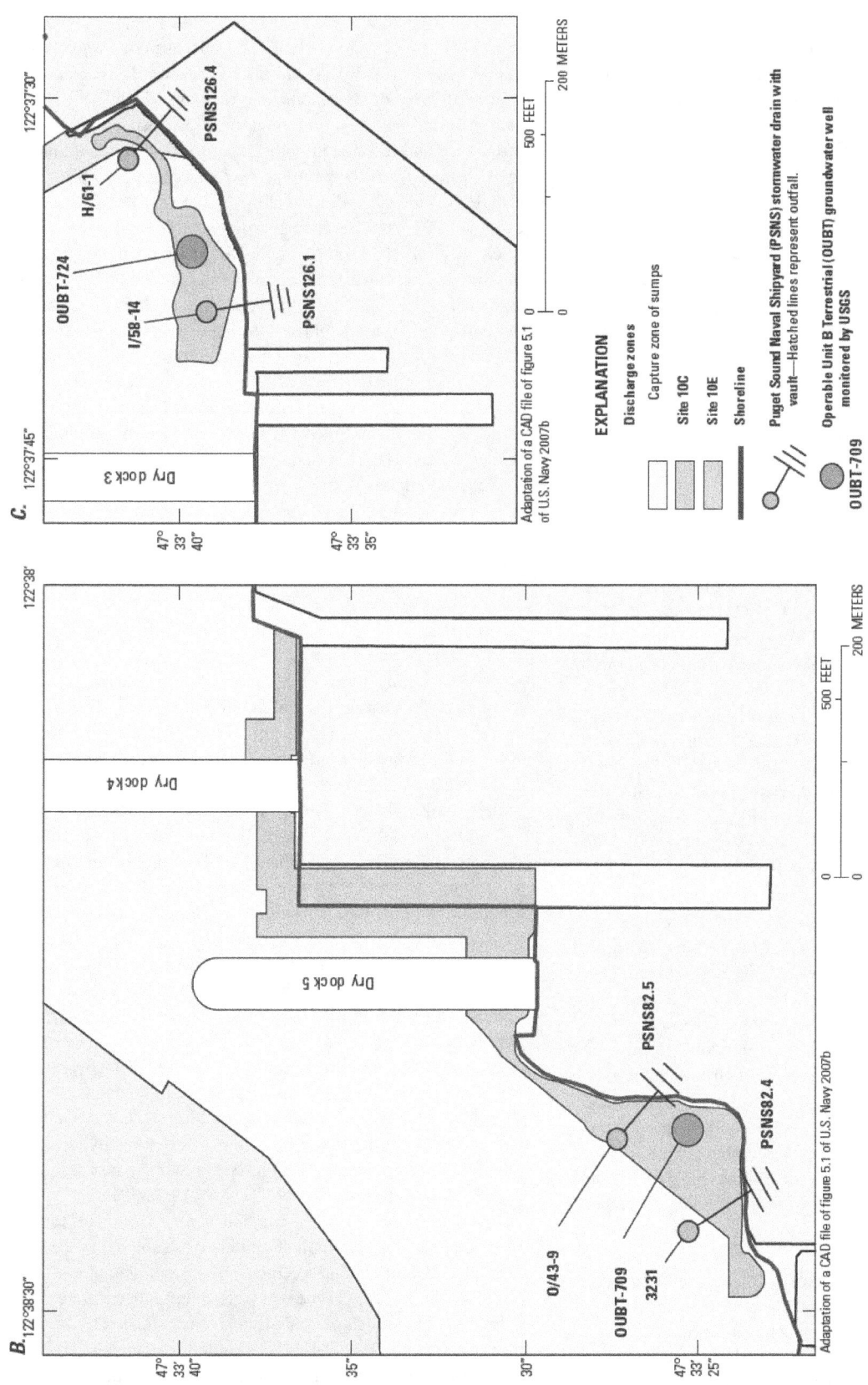

EXPLANATION

Discharge zones

☐ Capture zone of sumps

☐ Site 10C

☐ Site 10E

— Shoreline

⚬╱╱╱ **Puget Sound Naval Shipyard (PSNS) stormwater drain with vault**—Hatched lines represent outfall.

● **Operable Unit B Terrestrial (OUBT) groundwater well monitored by USGS**

OUBT-709

C.

PSNS126.4

H/61-1

OUBT-724

I/58-14

PSNS126.1

Dry dock 3

200 METERS

500 FEET

Adaptation of a CAD file of figure 5.1 of U.S. Navy 2007b

B.

Dry dock4

Dry dock 5

PSNS82.5

PSNS82.4

0/43-9

OUBT-709

3231

200 METERS

500 FEET

Adaptation of a CAD file of figure 5.1 of U.S. Navy 2007b

Figure 3.–Continued

Landfill Areas.—Previous work (Paulson and others, 2012) suggests that contaminated soils in landfill areas can release THg to BNC groundwater that may eventually result in FTHg releases to Sinclair Inlet. Two fill areas in the Capture Zone of the Sumps (fig. 3A) are not contained by a seawall and could potentially release THg to Sinclair Inlet, especially during some low tide conditions when the Sinclair Inlet water level is below the water level of the zone of depression of the sumps. Well Operable Unit B Terrestrial (OUBT)-709 is in a fill area (Site 10C (central), U.S. Navy, 1992; fig 3A–B), and was not sampled by the LTMP. The FTHg concentration of 31 ng/L in January 2008, from well OUBT-709, was the highest measured by USGS, other than those measured from the well adjacent to Site 2 where high THg concentrations have been documented (Paulson and others, 2012). The THg concentration of solids sampled from well OUBT-709 also was high (46.8 mg/kg), but may have been be affected by bias from sequential sampling described for the 2008 dry dock samples (see "Dry Docks" in this section).

The area upland of well OUBT-724 was identified as a fill area (Site 10E (eastern), U.S. Navy, 1992; fig. 3A and 3C) and is near the eastern boundary PSNS&IMF and the Capture Zone of the Sumps (Prych, 1997). Four unfiltered samples analyzed by the LTMP from well OUBT-724 (named LTMP-5 by the U.S. Navy) contained WTHg concentrations between 230 and 5,240 ng/L. The sample with a WTHg concentration of 5,240 ng/L was collected on October 18, 2005 at a tidal stage higher than other LTMP-5 samples. Groundwater from LTMP-5 collected in 2008 and 2009 contained the third highest WTHg concentrations (41.6 and 27.5 ng/L, respectively) of the LTMP for the respective year. FTHg concentrations less than 10 ng/L and PTHg concentrations less than 2 ng/L were measured from this well by USGS in 2008.

Dry Docks.—Water levels in the six dry docks on the BNC (fig. 3A) are kept below Sinclair Inlet water levels by sump pumps. Water from the sump wells is discharged by two systems: (1) sump well and process water from a combined collection system of Dry Docks 1–5 (fig. 3A) which are discharged alternately by pumps 4 and 5, and (2) the sump well and process water of the largest Dry Dock 6 (fig. 3A) discharged by pump 6. Cells 63 and 59 in Sinclair Inlet are the nearest cells to the discharge areas of sump well and process water from pumps 4 and 5, while cell 43 receives discharge from pump 6 (fig. 3A). The regular pumping of the sumps results in a groundwater zone of depression (Prych, 1997) defined as the Capture Zone of the Sumps (fig. 3A). Most of the THg released to the groundwater in the Capture Zone of the Sumps is expected to be accounted for as THg in the dry dock discharge.

Median FTHg concentrations of the combined Dry Dock 1–5 system and Dry Dock 6 in 2008 were 1.36 and 1.8 ng/L, respectively, and are not the subject of this report.

In 2008, THg concentrations of solids discharged from the dry docks relief discharge systems ranged from 1.27 to 5.17 mg/kg when TSS concentrations were greater than 0.5 mg/L. In contrast, THg concentrations of solids ranged from 8.74 to 17.7 mg/kg when TSS concentrations were less than 0.5 mg/L. In 2008, samples for PTHg were collected shortly after the dry dock pumps turned on, which likely caused solids near the pump inlet to be resuspended by turbulence and resulted in high TSS and PTHg concentrations at the beginning of the pumping cycle. Because the TSS sample was always collected at the end of each sampling event in 2008, the TSS concentrations in this last sample were likely different than samples collected at the beginning of the pumping cycle. If the population of solids in the PTHg sample collected at the beginning of the pumping cycle is different than the population of solids in the TSS sample collected at the end of the pumping, the THg concentration of dry dock solids (THg = PTHg / TSS) will be biased by sequential sampling during changing TSS conditions. THg concentrations of dry dock solids in this report were collected in a manner that precluded bias from sequential sampling of PTHg and TSS measurements.

Methods

Sampling

All sampling equipment and bottles were cleaned at the Wisconsin Mercury Research Laboratory (WMRL) as described by Lewis and Brigham (2004), except for the churn splitters that were cleaned in room-temperature 10 percent Trace-Metal Grade hydrochloric acid (HCl, volume to volume with deionized water) after removing metal-containing spigot parts. Details of the routine sampling methods by USGS are described in Huffman and others (2012), and only significant sampling deviations for the specific purposes of this study are described in detail. All samples were stored on ice and transported to the laboratory for processing.

Creeks.—Four creeks in Sinclair Inlet Basin (fig. 2) were sampled twice during the wet season of 2012 (February and March). Environmental parameters including pH, water temperature, dissolved oxygen, and turbidity were measured in each creek using a multiparameter sonde (YSI™ Incorporated, Yellow Springs, Ohio) and flow was measured using a FlowTracker (SonTek™, San Diego, California). Water was pumped from the centroid of the creeks into polyethylene terephthalate copolyester (PETG) bottles using a peristaltic pump for the determination of FTHg, PTHg, and TSS.

Steam Plant.—After conversion of the steam plant to RO processes, the steam plant effluent was sampled twice in 2011 (November and December). A 1.3-cm diameter perfluoroalkoxy (PFA) tube was passed through the grate at the top of the wet well and effluent was collected by peristaltic pump into a 2-L PETG bottle for FTHg and PTHg (Huffman and others, 2012). Water also was pumped into a polypropylene (PP) churn splitter and split into bottles for PTHg and TSS measurements.

Landfill Areas.—To understand THg dynamics in groundwater in the two landfill areas of concern (fig. 3, Site 10C and Site 10E), a 19-hour tidal study was conducted at well OUBT-709 and well OUBT-724 in 2011. The sampling dates for these tidal studies were chosen specifically to represent worst-case scenarios for possible release of THg to Sinclair Inlet. During the maximum higher high tide of the year (king tide), seawater from Sinclair Inlet extends the farthest inland as in the previous year, and may contact soils that have not been in contact with seawater during that time. During the ebbing cycle to lower low tide, the groundwater flows to Sinclair Inlet, potentially carrying extracted THg. The tidal studies included hourly sampling for about 19 hours at an elevation just below the water level of the well and sampling at the bottom of the well during higher high tide, higher low tide and lower low tide.

In addition to the hydrologic connection between the Sinclair Inlet and BNC groundwater, previous work (Paulson and others, 2012) also suggests a possible connection between groundwater and the stormwater drain systems on the BNC. In order to examine a possible connection between groundwater and stormwater drains in these two fill areas, two stormwater vaults upland of each well were also sampled a single time during the tidal study.

(a) Well OUBT-709 (Site 10C). On November 27, 2011, one sampling assembly was lowered to the bottom of the screened interval of well OUBT-709 and a second assembly was lowered to just below the elevation of the water level. A sampling assembly consisted of PFA tubing connected to a polytetrafluoroethylene (PTFE) sampling port on one end and a short piece of flexible C-FLEX® tubing on the other end inserted into a peristaltic pump.

Groundwater from the bottom of the well was sampled at three discrete times on November 27–28: near higher high tide (07:57 Pacific Standard Time (PST)), higher low tide (13:37) and lower low tide (01:20, November 28). Water was first pumped through a peristaltic pump at 300 mL/min into a flow-through chamber equipped with a multi-parameter YSI™ sonde to measure salinity (as specific conductance) and water temperature. Whole water was then pumped into a 2-L PETG bottle for analysis of FTHg, PTHg, and TSS. The bottom port had fallen to the bottom of the well for the final sample (01:20, November 28) and data from this sample did not satisfy quality assurance criteria. During bottom well sampling, the pump for the upper sampling assembly was off, and other than the three sampling times, the pump for the bottom sampling assembly was off.

Sampling and monitoring of the groundwater just below the water level was continuous, except when the pump was off during bottom sampling or when the sampling assembly was removed from the well for water level measurements. Water levels in the well were measured periodically and the upper sampling assembly was lowered or raised in response to the fluctuating groundwater level in the well (fig. A1). At about hourly intervals from 07:15, November 27, to 00:35, November 28, groundwater just below the fluctuating water level was pumped through a Meissner cartridge filter into 500-mL PFA bottles to be analyzed for FTHg concentrations. The filter was rinsed with blank water provided by the WMRL between each sample and was used until replaced with a new filter after it clogged at mid-experiment. Blank water was then pumped through the sampling assembly and the clogged, rinsed Meissner cartridge filter to assess cross-contamination between sequential samples. During higher high tide (07:28), higher low tide (12:48), and lower low tide (01:15, November 28), whole water from just below the water surface in the well was pumped into PETG bottles for analyses for PTHg and TSS concentrations. When water level measurements or sample collection was not occurring, groundwater just below the water level was pumped (in a manner that kept the C-FLEX® tubing clean for the next sample) into a YSI™ sonde flow-through chamber to continuously record environmental parameters.

During the ebbing tide of the well tidal sampling of November 27–28, 2011, water from O/43-9 (system PSNS82.5) and catch basin 3231 (system PSNS82.4), the nearest stormwater vaults to well OUBT-709 (fig. 3B), was pumped through a sampling assembly into a 2-L PETG bottle for analysis of FTHg, PTHg, and TSS.

(b) Well OUBT-724 (Site 10E). On December 26–27, 2011, the sampling procedure used at well OUBT-709 was repeated at well OUBT-724. Water from the bottom of the screened interval was monitored for environmental parameters and samples were collected for analysis of FTHg, PTHg, and TSS during higher high tide (07:15), higher low tide (12:38) and lower low tide (00:05, December 27). Water levels were monitored (fig. A2) and hourly samples from just below the fluctuating water level were filtered through a Meissner cartridge filter for FTHg analyses from 07:06, December 26, to 00:05, December 27, 2011. Field parameters from the water-level port were measured in the flow-through chamber equipped with a YSI™ sonde when other measurements were not being made.

All vaults in the stormwater system (PSNS126.4) upland of OUBT-724 contained freshwater on December 26. Freshwater from the most seaward vault of PSNS126.4 (fig. 3C, H/61-1) was pumped into a 2-L PETG bottle for analysis of FTHg, PTHg, and TSS. The salinity values of water in catch basins to the west of the well near the shoreline were measured, and the vault nearest the well that contained seawater (fig. 3C, I/58-14 of PSNS126.1) was sampled for FTHg, PTHg, and TSS analysis.

Dry Docks.—Because the goal of the dry dock sump sampling was to obtain the most accurate measurement of THg concentration of solids, every effort was made to ensure that the population of solids collected on the quartz fiber filter (QFF) for measurement of PTHg was the same as the population of particles collected on the polycarbonate filter used for measurement of SSC. The water samples of dry dock discharges were taken directly from a sampling port on the discharge pipe downstream of the pump into a PTFE churn splitter using PFA tubing with a C-FLEX® tubing connector.

The pumps of the dry dock sumps are controlled by water elevation sensors, and thus, the timing of the pumping cycle will depend on hydrologic conditions. The Dry Dock 6 pump cycles on for approximately 12 minutes per hour, and this cycle continues 24 hours per day, seven days per week. While the pump was on, water was collected from a sampling port on the discharge pipe at the top of the vertical riser (near the discharge pump on the lower level of the dry dock). Discharge water was collected sequentially into three churn splitters (for 14 minutes in November 2011) and four churn splitters (for 12 minutes in December 2011). Two acid-cleaned PTFE churn splitters were used for each dry dock pumping cycle and rinsed with Purelab® water between each use. While the water was churned, aliquots from each churn splitter were collected in a 1-L PETG bottle for PTHg and a 500-mL high-density polyethylene (HDPE) bottle for SSC analysis.

The Dry Dock 4 pump, which discharges sump water from the Dry Dock 1–5 system, cycles on for approximately 1 hour every 3 hours. This cycle continues throughout the day, seven days per week. While the pump was on, water was collected from a sampling port at the bottom of the vertical riser. In November 2011, three samples of discharge water were collected: immediately after the pump turned on, in the middle of the cycle, and near the end of the cycle (42 minutes of total pumping time). In December 2011, pump 4 turned on prematurely as the sampling team arrived at the pump station, then manually turned off. After about 8 minutes, the pump was manually turned on again and three sequential samples were taken over 12 minutes, before the pump cycle was completed. Because the pumping cycle was not complete, the suspended solids load for the cycle was not calculated. In addition, both SSC and THg of solids may have been affected by a short quiescent period of 12 minutes compared to the normal quiescent period of 3 hours or more.

Laboratory Processing.—Water samples filtered in the field (that is, groundwater samples collected just below the well water level) were acidified at the Washington Water Science Center (WAWSC) laboratory within 6 hours of collection with 20 ml of 6 moles HCl per liter of water in a laminar flow hood. Whole water samples of creek water, steam plant effluent, groundwater, and dry dock discharges for analysis of FTHg and/or PTHg were processed at the WAWSC using PFA filtering towers attached to a vacuum desiccator placed in a laminar flow hood as described by Lewis and Brigham (2004). TSS was measured gravimetrically using 0.45 μm pore-size, 47-mm diameter Nuclepore® polycarbonate filters (Huffman and others, 2012). Field collection, laboratory processing, and analyses are described in detail in Huffman and others, 2012.

Quality Assurance

Churn Splitters.—Churn splitters were used to collect representative samples of PTHg and solids. Churn splitters previously had not been used to collect mercury in any USGS program, and each PTFE and polypropylene churn splitter was pre-tested for THg contamination. Water used to dilute acid, rinse, and test the churn splitters was from an Elga Purelab® point-of-service unit supplied with RO water from an Elga Centra™ unit (Woodridge, Illinois). During the time of pre-testing of the churn splitters, WTHg concentrations of Purelab® water ranged from <0.04 to 0.08 ng/L (table A1). Water from the churn splitter was collected through the spigot into a PFA bottle (WTHg) or filtered through a QFF in a PFA filtering tower for FTHg and PTHg analysis. FTHg concentrations in water mixed in a PTFE churn splitter ranged from 0.04 to 0.22 ng/L, which were deemed unacceptable for collection of low-level FTHg samples. Therefore, all samples for FTHg analyses for the steam plant effluent were collected in PETG bottles. WTHg concentrations of samples mixed in PP churns splitters also were examined and withdrawn from the top of the PP churn splitter in order to determine if the PP churn splitters could be lowered into stormwater vaults to collect samples. WTHg concentrations from PP churn splitters were highly variable (0.2–1.2 ng/L), indicating that churn splitters should only be used to collect FTHg samples only in highly contaminated sites.

PTHg concentrations of the Purelab® water, mixed in PTFE churn splitters and transferred through the PTFE spigot, were less than the detection limit (0.019–0.040 ng/L), which depended on the amount of water filtered. PTHg concentrations of Purelab® water, mixed in PP churn splitters and transferred through spring-loaded plastic spigots also were below the detection limit (0.025–0.030 ng/L).

Replicates and Blanks.—Quality assurance of sampling methods included collection of field blanks, field replicates, and samples for comparison of different filtering methods (table A1). Field filtering of blank water through tubing at a creek site resulted in a FTHg concentration of 0.08 ng/L. The field filtering of blank water through sampling port, tubing, and new Meissner cartridge filters at the two groundwater sites resulted in FTHg concentrations of 0.09 and 0.11 ng/L.

FTHg concentrations were measured in three sets of field groundwater replicates and one set of creek replicates. The relative percent difference (RPD) was less than 5 percent for the three groundwater replicates and was 10.7 percent for the one creek sample. For one set of sequential FTHg groundwater samples, water from the first sample was filtered through a Meissner filter and collected in PFA bottles in the field. For the second sample, raw water was stored in a PETG bottle, taken to the laboratory, and filtered through a QFF. The RPD of FTHg concentration in the field-filtered sample relative to the sample filtered through a QFF was 24 percent.

Replicate samples of PTHg from the steam plant effluent and from Anderson Creek were collected. The RPD of low-PTHg concentrations in the steam plant effluent, collected by different methods (churn splitter and PETG bottle), was 43 percent. For comparison, the RPD of the creek samples was 6.6 percent.

The reporting level for mass on the filter was 0.03 mg. The TSS reporting level depended on the volume of water filtered. For low-TSS concentration samples, a nominal volume of 1 L was used. The increase of mass of filters after processing blank water, pumped through cleaned field equipment, ranged from 0.07 to 0.13 mg. Replicate PTHg or solids samples could not be collected during the rapid processing of sequential sampling during the pump cycle of the dry docks. The increase of mass on a filter after processing blank water that had rinsed a used PTFE churn splitter was 0.13 mg, which represented 7 percent of the 2 mg (0.47 L of 4.169 mg/L water) in the last sample from Dry Dock 6. The RPD of five creek field duplicates ranged from 0.4 to 8.6 percent. Two of the three TSS replicates for the groundwater study were within this range, but the RPD of duplicate TSS concentrations from the catch basin 3231 (in Site 10C was 39 percent.

To examine the possible crossover contamination from the sequential use of the same sampling assembly and cartridge filter at the well sites, the sampling port, tubing, and cartridge filter used previously for the same well were rinsed with blank water before filtering blank water into a PETG for FTHg analysis. The FTHg concentration in filtered blank water and filtered through a cartridge filter used to collect groundwater from OUBT-709 was 5.33 ng/L. This concentration represented between 1 percent (highest concentration) and 23 percent (lowest concentration) of the measured FTHg concentrations at OUBT-709. The FTHg concentration in blank water filtered through a cartridge filter used to collect groundwater from OUBT-724 was 0.73 ng/L. This concentration represented between 7 percent (highest concentration) and 18 percent (lowest concentration) of the measured FTHg concentration measured at OUBT-724.

The quality assurance data for this project is acceptable and no blank corrections were made. The FTHg concentrations (0.08–0.11 ng/L) and PTHg concentrations (<0.072 and 0.03 ng/L) in blank water passing through clean equipment for individual samples were low. Some crossover of THg between samples occurred from the use of a single cartridge filter for multiple well samples, but only represented 1 percent of the highest concentration of 496 ng/L collected from well OUBT-709 and 7 percent of the highest concentration of 10.9 ng/L collected from well OUBT-724. The RPD of field duplicates of FTHg in stream and well water ranged from 1.5 to 10.7 percent, while the RPD of PTHg in one stream field duplicate was 6.6 percent. Higher RPDs (24 and 43 percent) were only observed when different sampling methods were used to collect duplicate field concentrations during changing conditions. The filters used to measure TSS gained little mass (0.13 mg) after processing water pumped through clean equipment, and the crossover contamination

from reusing rinsed churn splitters in the field was about 5 percent. The RPD of 9 of the 10 field replicates was less than 10 percent.

Filtered Total Mercury Concentrations and Loadings from Potential Sources

Creeks

FTHg concentrations in four creeks draining into Sinclair Inlet (fig. 2) during the wet season of 2012 (1.05–4.0 ng/L, table 1) were considerably higher (Kendall tau <0.001) than those sampled during the 2008 dry season (0.39–0.81 ng/L, Paulson and others, 2012), but lower than the median WTHg concentration of 4.75 ng/L reported for rainwater (Brandenberger and others, 2010). Unlike the fairly uniform FTHg concentrations in these four basins during the dry season, FTHg concentrations varied among creeks during the wet season (fig. 4).

The differences of FTHg concentrations in creeks, between dry and wet season, suggest that FTHg loadings to Sinclair Inlet from local drainage basins be calculated from flows for the two seasons as outlined by the ENVVEST Program (Skahill and LaHatte, 2007). Average wet-season flows of 0.93, 0.32, 0.72, and 0.018 m^3/s for Gorst, Anderson, Blackjack, and Annapolis Creeks, respectively, contribute 94 percent of the total wet-season flow to Sinclair Inlet (2.104 m^3/s, table A2). Applying the average of the two measured wet-season FTHg concentrations (table 1) of 1.11, 2.05, 2.67 and 3.33 ng/L to the respective individual stream flows yield FTHg loadings of 19.0, 12.1, 35.2 and 1.0 g/yr for the 7-month wet season, for Gorst, Anderson, Blackjack, and Annapolis Creeks, respectively (table A2). Although Annapolis Creek contained the highest FTHg concentration, its wet-season flow was low (0.018 m^3/s), yielding a FTHg loading of only 1.0 g/yr. The highest FTHg loading (35.2 g/yr) was contributed by Blackjack Creek because it had the combination of the second-highest flow and second-highest FTHg concentration of the four creeks. When the median wet-season FTHg concentration of the four measured creeks (2.28 ng/L) was applied to the wet-season flow of the ungaged area, the total wet-season FTHg loading from creeks draining to Sinclair Inlet was estimated as 72.5 g/yr (table A2). This calculation assumed the sampling events were representative of the entire wet season; however, additional storm-event based sampling may be needed to more accurately quantify annual loading from wet-weather creeks.

The 7-month wet season contributes 80 percent of the total annual creek surface water volume delivered to Sinclair Inlet. The median dry-season FTHg concentration of 0.57 ng/L for all four creeks was applied to the 20 percent of the volume of water delivered to Sinclair Inlet (0.75 m^3/s over 5 months) to yield a dry-season FTHg loading from creeks draining to Sinclair Inlet of 5.6 g/yr.

Table 1.   Total mercury concentration in filtered water and of particles in four creeks draining into Sinclair Inlet, Kitsap County, Washington, February and March 2012.

[**Abbreviations:** m^3/s, cubic meters per second; mg/L, milligrams per liter; µS/cm, microsiemens per centimeter at 25 degrees Celsius; C, degrees Celsius; ng/L, nanograms per liter; mg/kg, milligrams per kilogram]

Station name	Date	Sample time	Instantaneous discharge (m^3/s)	Dissolved oxygen (mg/L)	pH	Specific conductance (µS/cm)	Temperature (°C)	Filtered total mercury (ng/L)	Particulate total mercury (ng/L)	Total suspended solids (mg/L)	Total mercury of solids (mg/kg)
Gorst Creek	02-29-12	1200	0.82	12.4	7.5	100	5.2	1.17	0.97	9.75	0.10
	03-14-12	1015	1.10	12.5	7.4	90	5.3	1.05	0.71	5.26	0.13
Anderson Creek	02-29-12	1345	0.33	12.5	7.6	69	5.6	2.08	1.61	23.71	0.07
	03-14-12	1115	0.20	12.6	7.5	64	5.5	2.02	1.77	17.50	0.10
Blackjack Creek	02-29-12	1645	0.88	12.6	7.5	130	5.0	2.48	1.60	29.25	0.05
	03-14-12	1515	2.01	12.7	7.4	139	5.0	2.86	2.14	44.40	0.05
Annapolis Creek	02-29-12	1500	0.07	12.3	7.4	100	5.5	4.00	2.64	13.83	0.19
	03-14-12	1300	0.14	12.6	7.3	71	5.0	2.66	2.77	53.60	0.05

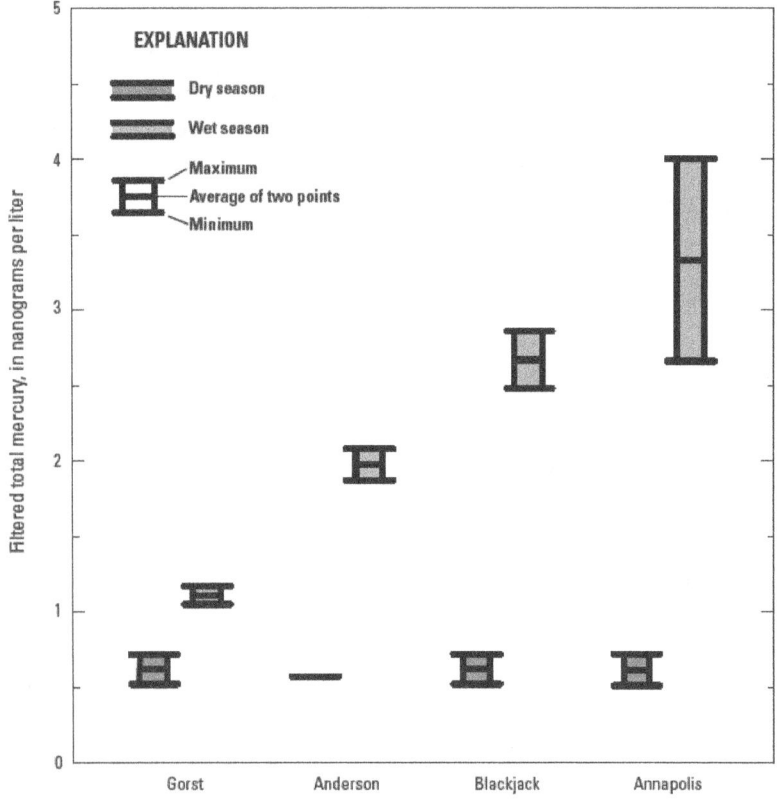

Figure 4.   Filtered total mercury concentrations in four creeks (Gorst, Anderson, Blackjack, and Annapolis Creeks) draining into Sinclair Inlet, during the 2008 dry season and the 2012 wet season, Kitsap County, Washington.

Steam Plant

Effluent from the steam plant, equipped with the new RO system, sampled in November and December 2011, had FTHg concentrations of 3.33 and 1.53 ng/L, respectively (table 2), which was much lower than concentrations before the process conversion (15.3–143 ng/L). The change in operation in the steam plant to a RO system and the diversion of the stack blow-down wash to the municipal wastewater treatment plant resulted in an average flow of water from the steam plant of 0.0019 m³/s in the autumn of 2011 (Mark Wicklein, U.S. Navy, written commun., February 2, 2012), which is a 20 percent reduction in flow from 2008 flows. The 20 percent reduction in flow and the 95 percent reduction in median FTHg concentrations from 2008 to 2011 resulted in a 97 percent reduction in the calculated FTHg loading to Sinclair Inlet from the steam plant (5.9 g/yr in 2009 to 0.15 g/yr in 2011).

Landfill Areas

Well OUBT-709 (Site 10C).—Well OUBT-709 is in the central fill area (10C, figs. 3A and 3B) and in the zone of depression of both dry dock systems. The two stormwater vaults nearest well OUBT-709 contained water with low conductivity (42 and 131 µS/cm) and low FTHg concentrations (2.85 and 1.63 ng/L) (table 3). These observations suggest that stormwater drains play no role in transporting saline water inland and all transport of saline water is through the intertidal region in the absence of a seawall.

On November 27, 2011, between 06:00 and 18:00 PST, the elevation of the water level at well OUBT-709 was between 2 and 3 m below the water elevation of the surface of Sinclair Inlet (fig. 5). Drawdown by the sumps of nearby Dry Docks 5 and 6 was probably responsible for lowering the water level in the well. During this time, groundwater was saline (salinity >26, see fig. A1) and was continuously flowing the short distance (30 m) from the Sinclair Inlet shoreline to the well and further upland. During the morning ebbing tide, FTHg concentrations just below the water level ranged from 32.1 to 50.2 ng/L (table 3). Near higher low tide at 12:36, FTHg concentrations began to increase, reaching a maximum of 126 ng/L at 14:10, FTHg concentrations then decreased to 22.9 ng/L and remained at relatively low levels until 19:00. Only after 20:50, when the elevation of Sinclair

Table 2.   Total mercury concentration in filtered water and of particles in the steam plant effluent on the Bremerton naval complex, Kitsap County, Washington, November and December 2011.

[**Abbreviations:** ng/L, nanograms per liter; mg/L, milligrams per liter; mg/kg, milligrams per kilogram; –, not measured; PETG, polyethylene terephthalate copolyester; PP, polypropylene]

Date	Time	Filtered total mercury (ng/L)	Particulate total mercury (ng/L)	Suspended-solids concentration (mg/L)	Total mercury of solids (mg/kg)	Collection device
11-16-11	1250	3.33	0.92	–	–	PETG bottle
11-16-11	1251	–	0.44	0.35	1.26	PP churn splitter
12-13-11	1001	1.53	0.10	–	–	PETG bottle
12-13-11	1011	–	0.07	0.19	0.36	PP churn splitter

Inlet declined below the well water level (+1.44 m MLLW), was groundwater flowing from the upland area to Sinclair Inlet. The FTHg concentration in this draining water was 228 ng/L at 21:00 and reached a maximum concentration of 496 ng/L at 22:12. By lower low tide at 00:27 (November 28), FTHg concentrations had decreased to 81.7 ng/L. FTHg concentrations in groundwater from the bottom of the well (table 3) were less than concentrations in groundwater from just below the water level sampled near the same time.

High FTHg concentrations in groundwater only were measured when groundwater was flowing from the upland area to Sinclair Inlet and indicates a source of FTHg upland of the well. The short duration of the release of FTHg suggests that chloride ions in the intruding seawater extracted THg from soils that would not have been in contact with seawater during normal tides. The release during the extremely high tides suggests that the upland soil releasing FTHg may have been near ground surface.

Well OUBT-724 (Site 10E).—Unlike well OUBT-709, the water elevation in well OUBT-724 responded quickly to the changes in water elevation of Sinclair Inlet during most of the tidal study (fig. 6). To some degree, this is likely because OUBT-724 is closer to the shoreline (10 m) and farther away from the nearest dry dock, and hence near the boundary of the dry dock drawn-down zone (Prych, 1997). During both ebbing tidal cycles, salinity decreased with the declining water elevations, indicating some amount of fresher water from upland areas was flowing past the well to Sinclair Inlet (fig. A2). FTHg concentrations in groundwater just below the water table in OUBT-724 were less than 5.5 ng/L in two-thirds of the samples. On December 26, during the ebbing tidal cycle to lower low tide (after about 19:00), the Sinclair Inlet water elevation declined below the well water level and water from the upland fill area was draining past well OUBT-724. During that time, FTHg concentration increased to a maximum of 10.8 ng/L (table 4).

Table 3. Field measurements and total mercury concentrations in filtered water and of particles in groundwater from well OUBT-709 and in stormwater in nearby drains on the Bremerton naval complex, Kitsap County, Washington, November 2011.

[**Abbreviations:** FTHg, filtered total mercury; PTHg, particulate total mercury; TSS, total suspended solids; THg, total mercury; mg/L, milligrams per liter; µS/cm, microsiemens per centimeter at 25 degrees Celsius; C, degrees Celsius; m, meters; bgs, below ground surface; ng/L, nanograms per liter; mg/kg, milligrams per kilogram; –, not measured]

Time	Dissolved oxygen (mg/L)	pH	Specific conductance (µS/cm)	Temperature (°C)	Sampling depth (m bgs)	FTHg (ng/L)	PTHg (ng/L)	TSS (mg/L)	THg of solids (mg/kg)
Stormwater vault O/43-9 (system PSNS82.5), sampled November 27, 2011									
1105	–	–	42	–	1.0	2.85	3.65	8.42	0.43
1107	–	–	40	–	2.8	–	–	–	–
Stormwater catch basin 3231 (system PSNS82.4), sampled November 27, 2011									
1148	–	–	131	–	0.8	1.63	3.21	8.33	0.39
1150	–	–	131	–	1.0	–	–	–	–
Well OUBT-709 groundwater samples collected just below the water level, November 27–28, 2011									
0727	–	–	–	–	4.3	50.2	–	–	–
0728	–	–	–	–	4.3	–	29.6	0.72	41.1
0830	1.6	6.9	41,700	12.6	4.3	40.6	–	–	–
0906	0.3	7.2	41,900	12.5	4.3	38.8	–	–	–
1007	0.9	7.3	41,100	12.3	3.7	35.4	–	–	–
1117	0.2	7.3	42,200	12.5	3.7	32.1	–	–	–
1201	0.3	7.3	41,800	12.7	3.5	42.4	–	–	–
1244	1.5	7.3	41,600	13.0	3.5	84.0	–	–	–
1248	2.4	7.3	41,800	12.9	3.5	–	39.5	1.37	28.9
1410	2.8	7.2	41,600	13.0	3.5	126	–	–	–
1515	0.4	7.4	42,500	11.6	3.5	33.9	–	–	–
1610	0.2	7.4	42,500	12.7	3.5	28.3	–	–	–
1726	0.1	7.4	42,500	12.3	3.5	26.6	–	–	–
1806	<0.1	7.5	42,700	12.6	3.5	25.1	–	–	–
1923	0.1	7.5	42,700	12.4	3.5	22.9	–	–	–
0035	0.4	7.4	42,600	12.9	3.7	73.2	–	–	–
2100	1.7	7.2	42,200	12.5	4.0	228	–	–	–
2212	4.3	7.1	41,500	13.7	4.0	496	–	–	–
2306	3.7	7.2	41,700	13.3	4.3	288	–	–	–
0027	3.4	7.2	41,900	13.6	4.4	81.7	–	–	–
0115	–	–	–	–	4.4	–	5.54	0.18	31.5
Well OUBT-709 groundwater samples collected near bottom of the well, November 27–28, 2011									
0757	1.6	6.9	41,700	12.6	6.7	40.4	26.6	0.78	34.1
1337	1.9	7.4	42,600	11.0	6.7	37.5	22.3	0.67	33.2
0119	4.0	7.3	41,800	11.7	9.1	65.7	[1]	[1]	34.3

[1] Before the 0119 sample, the sampling port dropped to the bottom of the well. The THg of solids is valid for the solids near the bottom, but the PTHg and TSS concentration of solids on which the THg concentrations is based are not representative.

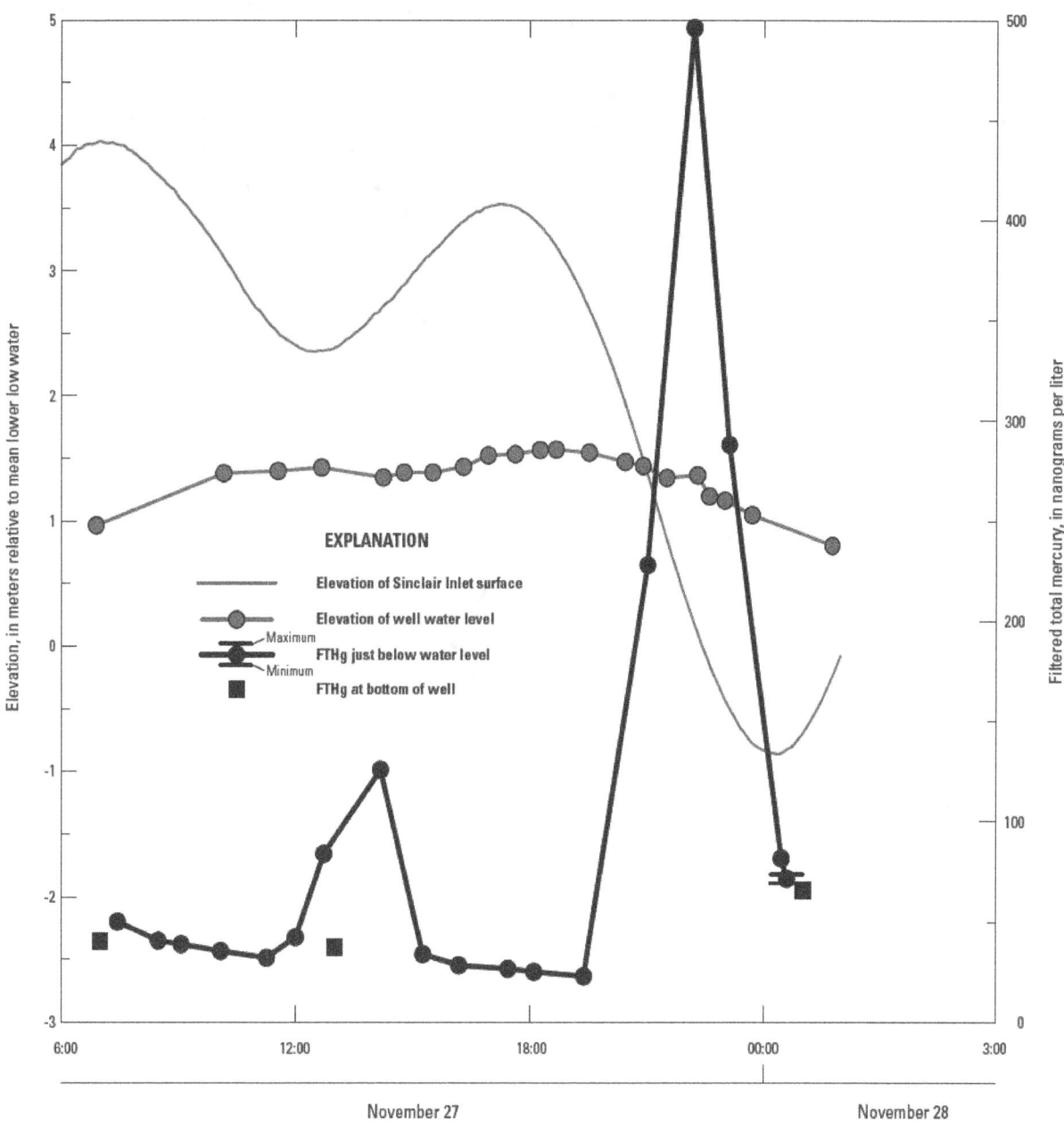

Figure 5. Water elevations in well OUBT-709 as compared to Sinclair Inlet and filtered total mercury concentrations in well water from OUBT-709, Kitsap County, Washington.

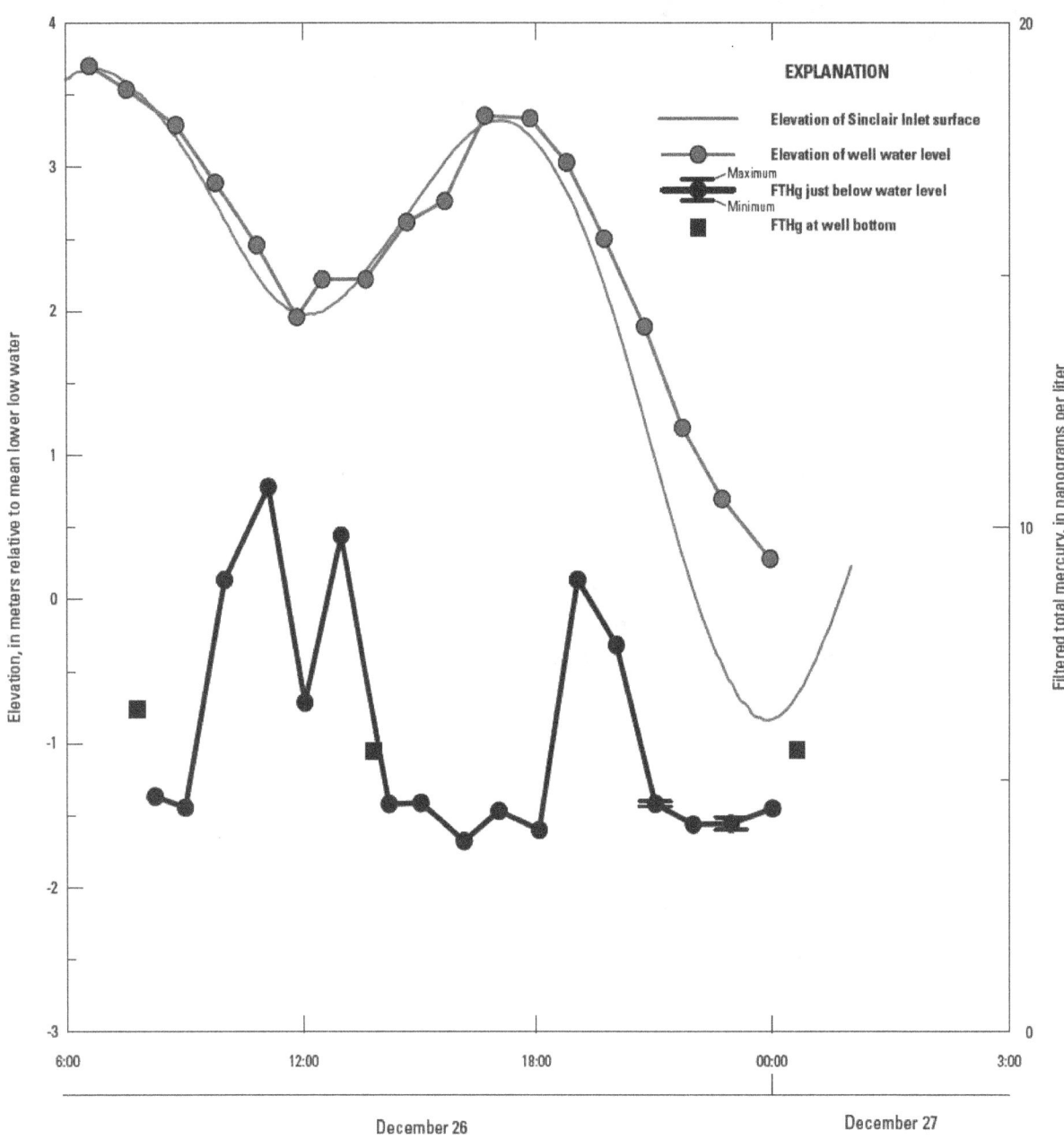

Figure 6.   Water elevations in well OUBT-724 as compared to Sinclair Inlet and filtered total mercury concentrations in well water from OUBT-724, Kitsap County, Washington, December 2011.

Table 4. Field measurements and total mercury concentrations in filtered water and of particles in groundwater from well OUBT-724 and in stormwater in nearby drains on the Bremerton naval complex, Kitsap County, Washington, December 2011.

[**Abbreviations:** FTHg, filtered total mercury; PTHg, particulate total mercury; TSS, total suspended solids; THg, total mercury; mg/L, milligrams per liter; µS/cm, microsiemens per centimeter at 25 degrees Celsius; C, degrees Celsius; m, meters; bgs, below ground surface; ng/L, nanograms per liter; mg/kg, milligrams per kilogram; –, not measured]

Time	Dissolved oxygen (mg/L)	pH	Specific conductance (µS/cm)	Temperature (°C)	Sampling depth (m bgs)	FTHg (ng/L)	PTHg (ng/L)	TSS (mg/L)	THg of solids (mg/kg)
				Stormwater vault H/61-1 near PSNS126.4, December 26, 2011					
1135	8.5	8.1	83	7.9	1.2	1.56	1.15	6.34	0.18
1136	8.4	8.0	78	7.9	2.0	–	–	–	–
				Stormwater vault I/58-14 (B-26) near PSNS126.1, December 26, 2011					
1110	6.3	7.5	41,200	10.1	1.8	1.58	3.44	4.22	0.82
1718	–	–	42,500	–	2.1	–	–	–	–
				Well OUBT-724 groundwater samples collected just below the water level, December 26–27, 2011					
0706	8.3	7.7	44,100	–	2.1	5.21	–	–	–
0715	–	–	–	–	2.1	–	1.02	0.34	2.96
0814	8.3	7.6	44,100	10.0	2.1	4.67	–	–	–
0900	7.9	7.7	44,300	10.3	2.1	4.45	–	–	–
1000	7.8	7.3	44,500	10.9	3.0	8.96	–	–	–
1107	7.8	7.2	40,000	10.5	3.4	10.8	–	–	–
1203	7.6	7.2	39,600	10.8	4.0	6.53	–	–	–
1238	–	–	–	–	4.0	–	0.93	0.56	1.67
1259	–	–	–	–	3.7	9.84	–	–	–
1415	7.0	7.6	43,400	10.6	3.7	4.52	–	–	–
1503	6.8	7.7	44,400	10.2	3.7	4.54	–	–	–
1610	7.2	7.7	44,500	10.3	3.0	3.78	–	–	–
1703	7.1	7.7	44,600	10.1	2.6	4.38	–	–	–
1805	7.1	7.7	44,700	9.4	2.9	4.00	–	–	–
1903	7.6	7.5	43,500	10.4	2.7	8.96	–	–	–
2002	7.3	7.5	43,800	10.4	3.4	7.67	–	–	–
2102	6.8	7.4	43,300	10.5	3.8	4.49	–	–	–
2103	–	–	–	–	3.8	4.56	–	–	–
2200	5.6	7.2	40,100	10.3	4.6	4.11	–	–	–
2258	5.6	7.3	39,900	10.0	5.2	4.04	–	–	–
2259	–	–	–	–	5.2	4.22	–	–	–
0001	6.8	7.2	35,500	9.5	5.3	4.43	–	–	–
0005	–	–	–	–	5.3	–	24.7	11.12	2.22
				Well OUBT-724 groundwater samples collected near bottom of well, December 26–27, 2011					
0747	7.2	7.5	44,800	10.1	5.5	6.39	6.70	4.12	1.62
1351	6.9	7.6	44,800	9.7	5.5	5.57	15.1	10.15	1.49
0038	–	–	–	–	5.5	5.58	1.71	1.67	1.02

FTHg concentrations in water from the two stormwater drains near OUBT-724 were 1.56 and 1.58 ng/L (table 4). These data suggest that FTHg concentrations in the well are not affected by the sampled stormwater drain systems.

In the previous estimate of FTHg loading to Sinclair Inlet (Paulson and others, 2012), release of FTHg by direct groundwater discharge from the Capture Zone of Sumps was thought to be minimal, however, release from fill areas Site 10C and Site 10E was identified as a data gap. In this study, we measured a maximum FTHg concentration of 10.8 ng/L in groundwater from well OUBT-724 in Site 10E (fig. 3), which suggests that the release of FTHg from this fill area is not as probable as earlier data had suggested. In contrast, FTHg concentrations in groundwater from a well located in a Site 10C (fig. 3) approached 500 ng/L suggesting the possible release of FTHg from this area. Additional intertidal studies will be required to determine the magnitude of FTHg loading from this fill area. The assumption of minimal groundwater release of FTHg from the Capture Zone of the Sumps to Sinclair Inlet needs to be re-evaluated.

The collection of stormwater samples near the monitoring wells in 2011 increased the total number of samples collected for measuring FTHg in PSNS&IMF fresh stormwater from two to five. The median FTHg concentration for the five samples was 1.63 ng/L, compared to the median value of 1.83 ng/L for the 2008 data alone. However, the maximum FTHg increased from 2.16 ng/L for the 2008 data set to 2.85 ng/L for the combined 2008 and 2011 data set. Based on these five values, the median loading of FTHg from PSNS&IMF stormwater to Sinclair Inlet is 0.43 g/yr with a range from 0.39 to 0.75 g/yr.

Dry Docks

The discharge of FTHg from the dry docks was estimated in Paulson and others (2012) at 18.3 grams per year and was not addressed in this document.

Revised Ranking of Potential Sources to Sinclair Inlet

Paulson and others (2012) ranked the sources of FTHg to Sinclair Inlet into four broad categories based on their estimated loading (table 5): I (less than 1 gram per year), II (single digit grams per year), III (tens of grams per year),

and IV (hundreds of grams per year). The two Category IV sources (stormwater drain PSNS015 and diffusion from Sinclair Inlet marine sediments) were not addressed in this report. This report provides a more accurate ranking of sources of FTHg to Sinclair Inlet (fig. 7 and table 5) and allows environmental managers to better prioritize future studies and clean-up actions.

The revised value for the total annual loading from creeks of 78.1 g/yr (72.5 g/yr during the wet season +5.6 g/yr during the dry season) is considerably larger than original the estimate of 27.1 g/yr (Paulson and others, 2012). Though this estimate is based on only two dry-season and two wet-season samples, the estimate is sufficiently robust to place creek discharges at the top of Category III sources higher than direct atmospheric discharge, greater Sinclair Inlet stormwater basins, and FTHg discharges from the BNC dry docks.

The conversion of the steam plant system from an ion-exchange system to a RO system reduced estimates of FTHg loading from a Category II source in 2008 (5.9 g/yr) to a Category I source in 2011 (0.15 g/yr). All known releases of FTHg from the Zone of Direct Discharge, since the conversion of the steam plant, were estimated to be less than 1 g/yr.

The 2011 groundwater studies from fill areas within the Capture Zone of the Sumps under worst-case tidal conditions suggest that the possibility of significant releases of FTHg from a fill area of interest on the eastern boundary of BNC (Site 10E) is not as likely as previously thought. In contrast, the FTHg release from a fill area in central PSNS&IMF (Site 10C), previously not studied in detail, may be substantial and may be contributing to the FTHg loadings from the dry docks. Groundwater flow data were not available to estimate the associated FTHg loading to Sinclair Inlet from the Capture Zone of the Sumps. Additional data from stormwater within the Capture Zone of the Sumps indicated an annual loading of 0.43 g/yr.

Based on the revised loading estimates of individual sources, the revised total loading of FTHg to Sinclair Inlet is 275 + 100s g/yr (table 5). This value is the sum of four main sources: (1) atmospheric wet deposition, unchanged at 34.6 g/yr; (2) greater Sinclair Inlet sources, increased to 99.9 g/yr owing to the wet season creek sampling; (3) BNC sources, decreased to 140 g/yr because of steam plant changes and an unknown amount from groundwater in the Capture Zone of the Sumps; and (4) diffusion from marine sediments, still estimated in the 100s g/yr.

Table 5.   Revised concentrations and loadings of filtered total mercury (FTHg) for Sinclair Inlet, Kitsap County, Washington.

[**Category:** I (less than 1 gram per year), II (single digit grams per year), III (tens of grams per year), and IV (hundreds of grams per year). **Abbreviations:** km^2, square kilometers; m^3/s, cubic meters per second; ng/L, nanograms per liter; g/yr, grams per year; –, not determined; NA, not applicable; PSNS, Puget Sound Naval Shipyard; IMF, Intermediate Maintenance Facility. Sources in italics were revised from Paulson and others, 2012.]

Source	Area (km^2)	Flow (m^3/s)	Number of samples	FTHg concentration (ng/L)		FTHg loading (g/yr)		Category
				Median	Range	Median	Range	
Exchange from Puget Sound, by difference						**420**	200–800	NA
Advection from Puget Sound	NA	98	40	0.20	<0.1–0.6	620	–	NA
Advection to Puget Sound	NA	100	–	0.33	–	1,040	–	NA
Direct atmospheric deposition	8.37	0.31	–	4.75	2.16 – 11.3	**34.6**	–	III
Greater Sinclair Inlet	86.3	3.17	–	–	–	**99.9**	–	NA
Creeks	74.0	–	–	–	–	78.1	–	III
Dry Weather (5 months)	–	0.75	10	0.57	0.39–0.81	5.6	3.8–8.0	–
Wet Weather (7 months)	–	2.10	8	2.28	1.05–4.0	72.5	1.1–35.2	–
Stormwater	12.3	0.093	3	4.25	4.02–4.62	12.3	11.6–13.4	III
Municipal effluent	NA	0.22	3	1.38	1.06–1.55	9.5	7.3–10.7	II
Bremerton naval complex	1.52	0.43	NA	NA	NA	**140+Capture Zone GW**	–	NA
Zone of direct discharge	0.10	0.009	NA	NA	NA	0.99	–	NA
Groundwater	NA	0.006	10	[1]4.06	[2]1.4–6	[1]0.81	[2]0.3–1.2	I
Stormwater	NA	0.001	0	[3]1.83	1.5–2.16	0.03	0.03–0.04	I
Steam plant	NA	0.002	2	[4]2.43	1.53–3.33	0.15	–	I
Vicinity of site 2	0.43	0.049	NA	NA	NA	120.2	–	NA
Groundwater	NA	<0.001	5	194	72–702	1.2	0.6–2.5	II
PSNS015 stormwater drain	NA	0.049	NA	NA	NA	119	–	IV
Freshwater	NA	0.010	1	144	NA	46	–	NA
Tidal flushing	NA	0.039	1	[4]58.8	NA	73	–	NA
Capture zone of sumps	0.99	0.368	NA	NA	NA	18.7+GW	–	NA
PSNS&IMF stormwater	NA	0.008	5	1.63	1.5–2.85	0.43	0.39–0.75	I
Dry docks	NA	0.36	NA	NA	NA	18.3	–	III
Dry Docks 1–5	NA	0.16	7	[2]1.36	[2]0.63–4.16	6.9	[2]3.2–26	NA
Dry Dock 6	NA	0.20	4	[2]1.81	[2]0.97–2.38	11.4	[2]8.2–13.9	NA
Groundwater	NA	unknown	10	2.62	1.07–31	Significant release possible from site 10C and unlikely from 10E		
Diffusion from marine sediments	8.37	NA	24	9.30	5.07 – 25.2	**100s**	100s	IV
TOTAL SINCLAIR INLET SOURCES						**275 + 100s**		

[1] Using concentrations of filtered total mercury from Wisconsin Mercury Research Laboratory.

[2] Range calculated from median by setting non–detectable values from the National Water Quality Laboratory to 0 and to 6 ng/L.

[3] Using FTHg concentrations from PSNS&IMF stormwater.

[4] Flux–weighted average during the March 31, 2010 ebb cycle.

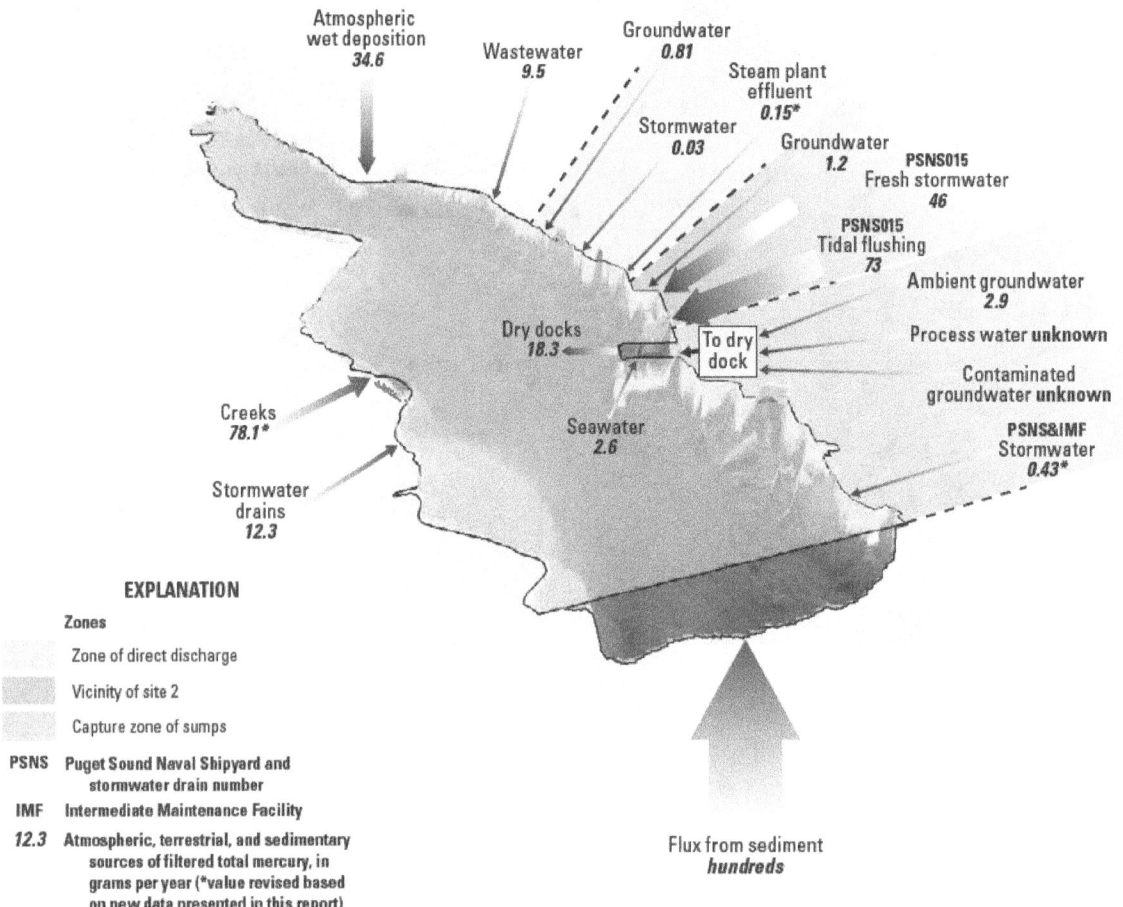

Figure 7. Revision of loadings of filtered total mercury from freshwater sources, from seawater recycled in and out of Bremerton naval complex, and from net advective transfers between Sinclair Inlet and Puget Sound (Paulson and others, 2012), Kitsap County, Washington.

Total Mercury Concentrations of Solids and Solids Loadings from Potential Sources

Creeks

TSS concentrations in the four creeks during the wet season ranged from 5.26 to 44.4 mg/L, excluding an outlier of 53.6 mg/L (table 1). The median TSS concentration in creeks sampled during wet season flow (17.5 mg/L) was higher (Kendall tau <0.001) than the median TSS concentration measured during dry weather base flow conditions in 2008 (2.4 mg/L). THg concentrations of creek solids collected during the wet season ranged from 0.05 to 0.19 mg/kg (table 1), and were not significantly different (Kendall tau = 0.11) than dry season THg concentration of solids.

THg concentrations of solids suspended in creeks collected during wet season were lower than concentrations on solids suspended in the water column of Sinclair Inlet (median of 0.23 mg/kg, Paulson and others, 2012). Since estimating annual loads of solids from watersheds requires an extensive data set and calibrated model, the small amount of data collected in this study does not warrant such an estimate.

Steam Plant

THg concentrations of steam plant effluent solids were 1.26 and 0.36 mg/kg in November and December 2011, respectively (table 2). The average THg concentrations of solids in 2011 (0.81 mg/kg) is a 95 percent reduction compared to the average concentration in 2008 (17.1 mg/kg. range: 2.95–68.7 mg/kg), when the steam plant utilized an ion-exchange demineralizing process.

The average SSC of two samples from the steam plant in 2011 was 0.26 mg/L, which was about a 75 percent reduction from median TSS concentrations measured in 2008. With the 20 percent reduction in flow, the estimated loading of solids from the steam plant in 2011 was 0.017 metric tons per year, which is about an 80 percent reduction in solids loading from 2008 estimates (0.09 metric tons per year).

Landfill Areas

THg concentrations of suspended solids in three samples collected from just below the water level and from three samples collected near the bottom of the well in Well OUBT-709 (fig. 3B, Site 10C) ranged from 28.9 to 41.1 mg/kg and from 33.2 to 34.3 mg/kg, respectively (table 3). In contrast, the solids collected from the nearby stormwater drains contained lower THg concentrations of solids (0.39 and 0.43 mg/kg).

At Well OUBT-724 (fig. 3B, Site 10E), THg concentrations of solids ranged from 1.67 to 2.96 mg/kg near the water table and ranged from 1.02 to 1.62 mg/kg near the bottom of the well (table 4).The THg concentration of solids from a nearby stormwater drain containing freshwater (fig. 3C, H/61-1) was 0.18 mg/kg (table 4). THg concentration of solids of the stormwater drain containing saline water (fig. 3C, I/58-14) was higher (0.82 mg/kg) and similar to the median concentrations of OU B Marine sediment (Paulson and others, 2012).

The selected groundwater samples analyzed for TSS and PTHg during the fill area tidal studies suggest that there are elevated levels of THg of solids in the groundwater in the central fill area. Additional intertidal studies will be required to determine the magnitude of THg of solids loading from the central fill area.

Dry Docks

During three of the four dry dock pumping cycles that were sampled in November and December 2011, SSC decreased with time (table 6, fig. 8). At the end of the pumping cycle from Dry Dock 6 (pump 6) in December, the SSC increased owing to the presence of large, buoyant, dark-colored clusters. These data suggest that SSC can change dramatically during the pumping cycle and the THg concentrations of suspended solids in dry dock discharge water calculated in 2008 could have been biased because of sequential sampling during changing TSS conditions.

The method of collecting dry dock discharge water into PTFE churns, splitting samples to PETG bottles for PTHg and HDPE bottles for SSC while churning, and then

Table 6. Total mercury concentration of particles in dry dock discharges on the Bremerton naval complex, Kitsap County, Washington, November and December 2011.

[All samples were collected from a polytetrafluoroethylene (PTFE) churn splitter. **Abbreviations:** PTHg, particulate total mercury; SSC, suspended-solids concentration; THg, total mercury; ng/L, nanograms per liter; mg/L, milligrams per liter; mg/kg, milligrams per kilogram]

Date	Time	PTHg (ng/L)	SSC (mg/L)	THg of solids (mg/kg)
Dry Docks 1–5 through pump 4 (41.5 mins, total discharge = 1,254 m³, solids loading = 8.57 kilograms)				
11-16-11	1350	13.5	0.74	18.3
11-16-11	1354	5.48	0.60	9.09
11-16-11	1357	6.36	0.72	8.86
11-16-11	1411	7.59	1.18	6.41
11-16-11	1425	6.58	0.77	8.60
Dry Docks 1–5 through pump 4 (incomplete pumping cycle)				
12-13-11	1238	17.6	1.73	10.2
12-13-11	1241	5.56	0.54	10.2
12-13-11	1246	5.20	0.39	13.4
Dry Dock 6 (14 mins, total discharge = 728 m³, solids loading = 2.95 kilograms)				
11-16-11	1102	5.63	2.68	2.10
11-16-11	1105	3.43	0.54	6.31
11-16-11	1110	3.40	0.26	13.0
Dry Dock 6 (11 mins, total discharge = 624 m³, solids loading = 2.18 kilograms)				
12-13-11	1137	11.5	4.78	2.40
12-13-11	1139	5.69	0.45	12.8
12-13-11	1143	2.43	0.39	6.16
12-13-11	1146	5.28	4.17	1.27

filtering the entire contents of each PETG and HDPE bottle, precluded temporal bias and assured that the populations in the corresponding PETG and HDPE bottles were similar. Thus, this method allowed an unbiased calculation of THg concentration of solids. There were considerable variations of THg concentrations of solids during each of the pumping cycles (table 6, fig. 8), with THg concentrations of solids generally higher in samples with lower SSC concentrations. For example, the sample with the second highest THg concentration of solids (13.4 mg/kg) contained the lowest SSC concentration (0.39 mg/L). In contrast, the sample with the lowest THg concentration of solids (1.27 mg/kg) contained the second- highest SSC concentration (4.17 mg/L).

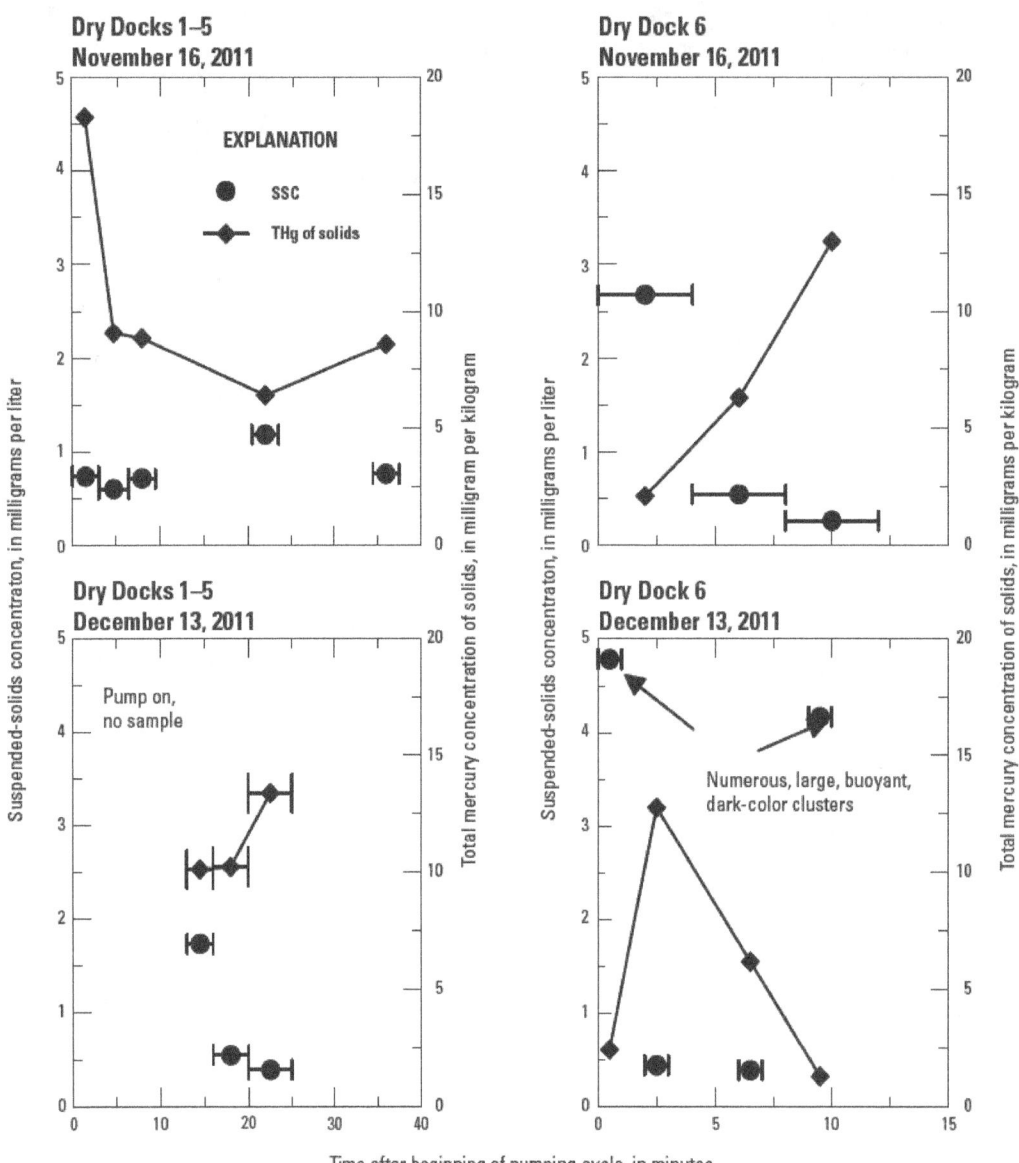

Figure 8. Suspended-solids concentrations (SSC) and total mercury (THg) concentration of solids during pump cycles of Dry Docks 1–5 and 6 (pumps 4 and 6, respectively) at the Bremerton naval complex, Sinclair Inlet, Kitsap County, Washington, November and December 2011.

The SSC in the dry dock discharges were integrated over the three complete pumping cycles to estimate solids discharge per pumping cycle. The loading of suspended solids for the November and December cycles for pump 6 were 2.95 and 2.18 kg of solids per pumping cycle (table 6), respectively. These per-cycle solids loadings were scaled up to annual solids loadings using pumping cycles that were consistent with annual average flows of Paulson and others (2012). Using a nominal pumping rate of 12 minutes every hour for pump 6 (Bruce Beckwith, U.S. Navy, written commun., May 6, 2008), the average load of 2.56 kg per pumping cycle yields an annual solids loading from Dry Dock 6 of 16 metric tons of solids. The solids loading from the November 2011 pumping cycle of Dry Docks 1–5 (pump 4) was 8.57 kg per pumping cycle. Using a nominal pumping cycle of one hour every three hours for the Dry Docks 1–5 system, yields an annual loading of 30 metric tons of solids. Even though the median solids concentration in 2008 (0.85 mg/L) was higher than the median solids concentration in 2011 (0.66 mg/L), the sampling in 2008 missed the beginning of the pumping cycle and higher solids concentrations were observed in 2011. The integration of the solids concentration over the entire pumping cycle provides a more accurate estimate of the actual solids loading for each pumping cycle. If more accurate estimates of solids loadings from the dry docks are needed in the future, more measurements of per-cycle solids loadings over a variety of operating conditions are needed.

When all the data for THg concentrations of solids suspended in the dry dock discharge water are plotted against SSC, THg concentrations of suspended solids in dry dock discharge water decrease asymptotically with increasing SSC (fig. 9), except for the initial sample from Dry Docks 1–5 during both sampling events. This may be partially explained by settling of the coarser fraction of the suspended solids in the water column of the sump well during the quiescent period between pumping cycles. With time, the SSC will decrease and the remaining solids in the water column will be smaller sized. Concentrations of metals of solids increase with decreasing particle size because of increasing surface-to-volume ratios. When relatively undisturbed sump water is pumped, SSC will be low and THg of the fine-grained suspended solids will likely be high. Occasionally, coarse solids containing lower THg concentrations may

be resuspended resulting in higher SSC and lower THg concentrations of solids. The resuspension of coarser solids from the bottom of the sump well can occur at the beginning of the pumping cycle when the water column is disturbed shortly after the pumps are turned on, or at the end of the cycle when the water level is low and the pumped water includes sediment at the bottom of the sump well. Higher SSC and lower THg of solids were measured at the beginning and end of the December 2011 sampling of the Dry Dock 6. THg concentrations of these two suspended solids samples were low, and similar to those found in Sinclair Inlet sediment (0.9 mg/kg, Paulson and others, 2012).

The dry dock data (n = 13), excluding the two samples from the beginning of the Dry Dock 1–5 pumping cycle, was modeled using the mixture of two populations of suspended solids with fixed THg concentrations of solids. The THg concentration of smaller-sized suspended solids at low SSC was set to 13.4 mg/kg, the highest observed concentration of the data set that was modeled. The THg concentration of the coarser suspended solids with variable SSC was set at 0.9 mg/kg. This value is the depositional weighted average STHg concentration for OU B Marine between 2003 and 2007. This STHg average is higher than the arithmetic average because the deposition of fine-grained sediment in depositional areas (appendix B, Paulson and others, 2012) containing higher STHg concentrations (Paulson and others, 2010) are more highly weighted than coarser sediment in erosional areas containing lower STHg concentration. The data set was modeled using this mixture of these two types of suspended solids, with the SSC of the higher THg-concentration solids being the only unknown. A least-square non-linear regression model was used to solve for this SSC as 0.32 mg/L and this model is shown in figure 9. The constant discharge of 0.32 mg/L of 13.4 mg/kg suspended solids translates into an annual discharge of 1.4 metric tons per year from Dry Docks 1–5 and 1.8 metric tons per year from Dry Dock 6. The average THg concentrations for the two dry dock systems, weighted by the loadings of the two different populations of solids were 1.48 and 2.31 mg/kg, respectively. The average of 1.48 mg/kg for Dry Dock 1–5 should be considered a minimum because it does not account for the elevated THg concentrations of solids at the beginning of the pumping cycle.

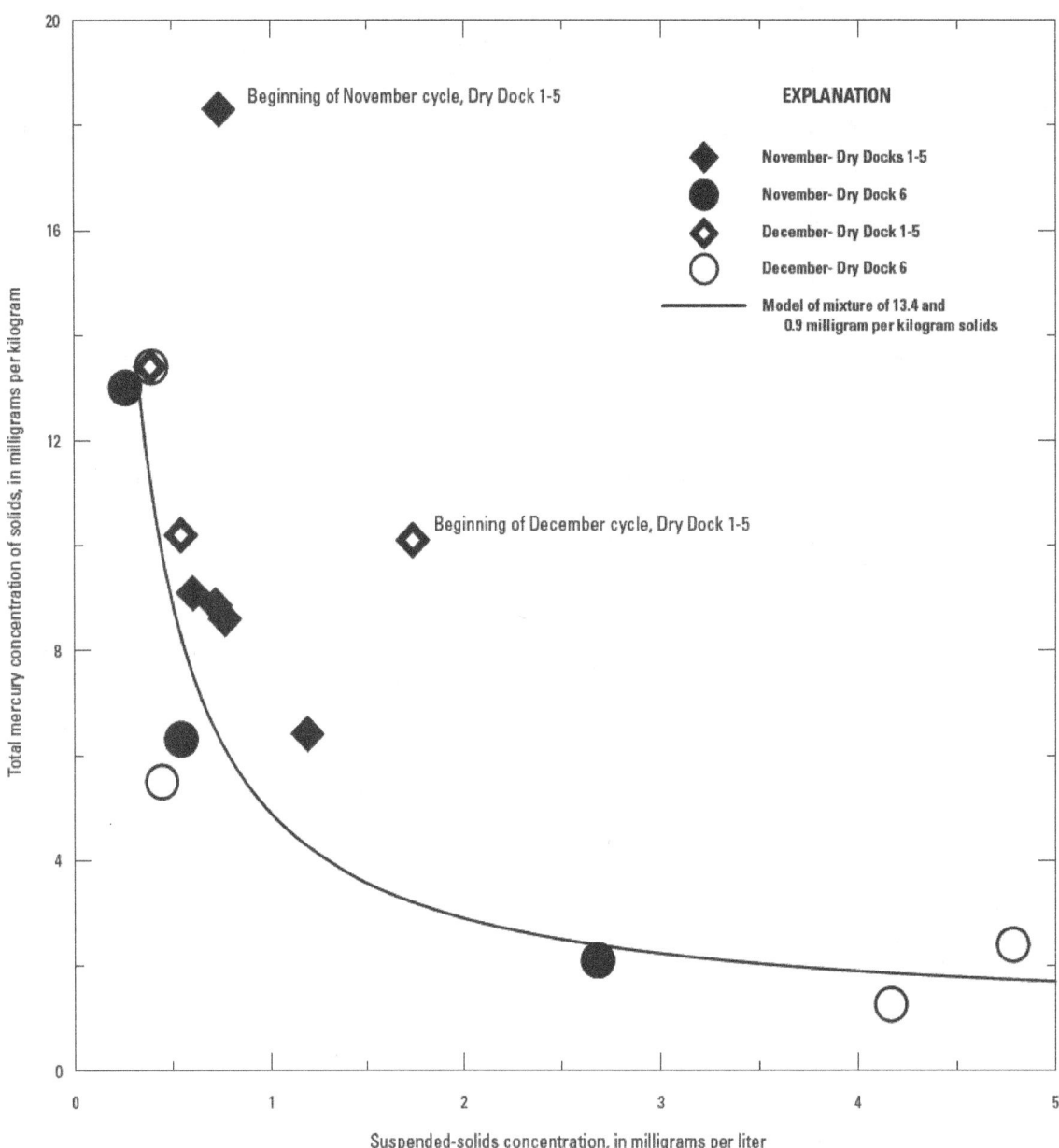

Figure 9.   Total mercury concentrations of solids versus suspended-solids concentration in dry dock discharges during four pumping cycles, Sinclair Inlet, Kitsap County, Washington, November and December 2011.

Paulson and others (2012) identified four possible sources of solids discharged by the pumps: (1) ambient groundwater containing suspended solids with a THg concentration of about 1.1 mg/kg, (2) suspended solids in marine water containing predominately resuspended sediment from OU B Marine (about 0.9 mg/kg), (3) solids from naval operations suspended in process water discharged in the sump well, and (4) solids suspended in fresh or saline groundwater transported through contaminated soils of the BNC. Paulson and others (2012) also calculated that about 70 percent of the water discharged by the pumps was seawater seeping into the dry docks. Therefore, the suspended solids containing 0.9 mg/kg THg likely originate from OU B Marine sediment that entered the dry dock systems with the seawater. Since solids suspended in ambient groundwater had relatively low THg concentrations, the source of solids containing 13.4 mg/kg THg (table 6) must have some combination of the latter two sources. Recent data of THg concentrations of solids suspended in well water from well OUBT-709, which is located between Dry Docks 5 and 6, indicated that solids transported to the dry docks through contaminated soils with groundwater is the most plausible source of the high-THg concentration solids observed suspended in dry dock discharge water. No data on THg concentrations of suspended solids in the dry dock process water are available to assess if the naval operations are a source of the observed high-THg concentration of solids. However, process water with high turbidity is typically diverted to the sewer system rather than to the dry dock sump wells (U.S. Navy, 2012). In November 2011, solids suspended in well water from well OUBT-709 contained THg concentrations that ranged from 28.9 to 41.1 mg/kg (n = 6), suggesting that this is the likely source of high THg solids in the dry dock system.

Potential Effects on Natural Attenuation of Sinclair Inlet Sediment

Previous investigations indicate that settling of suspended solids in the Sinclair Inlet water column is contributing to the recovery of OU B Marine sediments, though the decrease in THg of marine sediment in greater Sinclair Inlet between 2003 and 2010 is not yet statistically significant (Paulson and others, 2012). Data collected in 2011 and 2012 provided new data on THg concentrations of solids suspended in water discharged to Sinclair Inlet that contribute to the pool of solids suspended in Sinclair Inlet. Median THg concentrations of creek solids suspended in creek water during the wet season (0.05–0.19 mg/kg) were similar to those collected during dry weather and were lower than solids suspended in the water column of Sinclair Inlet (median concentration of 0.23 mg/kg). Conversion of the steam plant to RO system reduced THg concentrations of solids in effluent by 95 percent between 2008 and 2011 to a median concentration less than 1 mg/kg. It is unknown to what extent suspended solids in groundwater are transported to Sinclair Inlet. THg concentration of solids in groundwater collected at a fill area on the eastern boundary of BNC ranged between 1 and 3 mg/kg. The THg concentration of solids in groundwater collected from fill area in the center of BNC ranged from 29 to 41 mg/kg.

Discharges from four BNC sources contain suspended solids with THg concentrations higher than average STHg concentrations of OU B Marine, which may inhibit the decrease of STHg concentrations. Three of these four BNC sources were addressed in this report: steam plant effluent and the discharges from the two dry dock systems. Tidal flushing of stormwater drain PSNS015 (fig. 3A) is the fourth source of high-THg concentration solids and data on this source were not biased by sequential sampling (Paulson and others, 2012). In addition to having acquired unbiased THg concentrations of suspended solids from all four sources through the sampling methods presented in this report, recent sampling over the pumping cycles of the two dry dock systems provides more accurate estimates of the loadings of solids to OU B Marine.

Given these more accurate estimates of THg concentrations of solids and loading estimates of solids from these four sources, a broad assessment of the effect of these higher THg concentration solids on the recovery of STHg concentrations in OU B Marine can be attempted. Because the origins of the solids depositing to the surface sediment layer of OU B Marine are not known and the sediment concentration represents deposition, resuspension, and transformation activities over the last 20–40 years rather than recent deposition; only the incremental change in STHg concentration, owing to the solids discharged from BNC, is estimated. The net effect of high THg solids discharging from BNC on OU B Marine surface-sediment concentrations (mg/kg) is represented as:

$$\text{Net Effect} = \text{Concentration of settling solids} - \text{Sediment concentration} \tag{1a}$$

Substituting the formula for the weighted average of the concentration of settling solids:

$$\text{Net Effect} = -C_{sediment} + \frac{(C_{sources} \times L_{sources}) + (C_{wc} \times L_{wc})}{L_{sediment}} \tag{1b}$$

where

$C_{sediment}$ is THg concentrations of OU B Marine surface sediment (mg/kg);

$C_{sources}$ is THg concentrations of suspended solids from BNC sources (mg/kg);

C_{wc} is the average THg concentrations of suspended solids in the upper layer of OU B Marine water column (mg/kg);

$L_{sources}$ is the annual loading of suspended solids from BNC sources (metric tons/yr);

L_{wc} is the annual loading of suspended solids from the water column (metric tons/yr); and

$L_{sediment}$ is the annual sedimentation of solids over the area which the discharged solids are dispersed (metric tons/yr).

Given that $L_{sediment} = L_{sources} + L_{wc}$, we can multiply $C_{sediment}$ by $(L_{sources} + L_{wc})/L_{sediment}$, so that equation 1b becomes:

$$\text{Net Effect} = -C_{sediment} \times \frac{(L_{sources} + L_{wc})}{L_{sediment}} + C_{sources} \times \frac{L_{sources}}{L_{sediment}} + C_{wc} \times \frac{L_{wc}}{L_{sediment}} \tag{1c}$$

Rearranging to give:

$$\text{Net Effect} = -\left(C_{sediment} - C_{sources}\right)\left(\frac{L_{sources}}{L_{sediment}}\right) - \left(C_{sediment} - C_{wc}\right)\left(\frac{L_{wc}}{L_{sediment}}\right) \tag{1d}$$

Given that $L_{wc} = L_{sediment} - L_{sources,}$

$$\text{Net Effect} = \left(C_{sources} - C_{sediment}\right)\left(\frac{L_{sources}}{L_{sediment}}\right) + \left(C_{wc} - C_{sediment}\right)\left(\frac{L_{sediment} - L_{sources}}{L_{sediment}}\right) \tag{1e}$$

The second term on the right side of the equation represents the effect of deposition of suspended solids (mg/kg), or natural attenuation:

$$\text{Effect of Deposition of Suspended Solids}\left(\frac{mg}{kg}\right) = \frac{(C_{wc} - C_{sediment}) \times (L_{sediment} - L_{sources})}{(L_{sediment})} \tag{2}$$

The first term on the right side of the equation is the incremental change of STHg (mg/kg), or the inhibition to natural attenuation, in OU B Marine because of the high-THg concentration solids discharged from BNC sources:

$$\text{Incremental Change of STHg}\left(\frac{mg}{kg}\right) = \frac{(C_{sources} - C_{sediment}) \times L_{sources}}{L_{sediment}} \tag{3}$$

Therefore, equation 1 also can be stated as:

$$\text{Net Effect}\left(\frac{mg}{kg}\right) = \text{Incremental Change of STHg}\left(\frac{mg}{kg}\right) + \text{Effect of Deposition}\left(\frac{mg}{kg}\right) \tag{1f}$$

It may take years to decades for the settling of present-day suspended solids to be reflected in STHg concentrations in the surface layer of sediment. An accurate estimate of the timing of decreases of STHg concentrations requires a sediment model that defines the mixing of sediment with depth. With that caveat in mind, we can apply these equations by assuming an area in OU B Marine over which the solids from each source are deposited. Two scenarios that represent the range between minimal dispersion and maximum dispersion of these solids within OU B Marine were investigated. Identifying the actual degree of dispersion of these BNC solids between the minimal and maximum degree of dispersion requires complex modeling of the sediment transport and is beyond the scope of this study. In addition, the solids discharged by the BNC sources could be dispersed beyond OU B Marine.

Minimal Dispersion Scenario

In the worst-case scenario of minimal dispersion of these solids, we assume solids from the four sources deposit near their discharge points. Because of resolution of available data, deposition of these high THg-concentration solids was limited to a single 500-ft cell (23,225 m^2 each). For the

minimal dispersion scenario, $C_{sediment}$ and $L_{sediment}$ are specific to the cell in which the solids are discharged (table 7). In the sedimentation model (appendix B, Paulson and others, 2012), the sedimentation rate in each cell is dependent on the percentage of fines in the sediment. Similar to seven other cells in OU B Marine, no sedimentation is expected in cell 39 (fig. 3A), the receiving cell for the tidal flushing from PSNS015, because the percentage of fines in this cell is less than 50 percent. The other grid cells receiving solids from BNC sources (fig. 3A, cell 26 for the steam plant, cell 63 or 59 for pumps 4 and 5, and cell 43 for pump 6) contain sediment between 50 percent and 85 percent fines and were assigned a sedimentation rate of 0.09 gm cm^{-2} yr^{-1}, which translates to a $L_{sediment}$ of 20.9 metric tons per year for the 23,225 m^2 cells (table 7). Hypothetically concentrating the solids from BNC sources in these four cells under the minimal dispersion scenarios also means that the other 59 of 64 cells of OU B Marine, where solids are modeled to settle (percentage of fines greater than 50 percent, appendix B, Paulson and others, 2012), would not be affected by these sources and would more fully respond to the effects of the settling of present-day lower THg-concentration solids suspended in water column of OU B Marine.

Table 7.   Estimated net effect of high mercury solids discharged from Bremerton naval complex sources on Operable Unit (OU) B Marine sediment under two hypothetical dispersion scenarios.

[$C_{sources}$: concentration of total mercury of solids originating from four sources on the Bremerton Naval Complex (BNC). $L_{sources}$: loading of solids originating from four sources on the BNC. Cell: 23,225 m^2 area in OU B Marine to which BNC sediments are hypothesized to deposit (see appendix B in Paulson and others 2012). $C_{sediment}$: concentration of OU B Marine surface sediments, average of 2003, 2005, and 2007 from U.S. Navy 2006a, 2006b, 2008b, respectively. $L_{sediment}$: sedimentation rate, based on percentage of fines, Paulson and others, 2012. **Measured Change 2005–10:** from 2010 in U.S. Navy, 2011. **Abbreviations:** mg/kg, milligram per kilogram; MT/yr, metric tons per year; PSNS015, Puget Sound Naval Shipyard stormdrain outfall 015. See text for equations (eq.) 1f, 2, and 3]

Sources	$C_{sources}$ (mg/kg)	$L_{sources}$ (MT/yr)	Cell	$C_{sediment}$ (mg/kg)	$L_{sediment}$ (MT/yr)	Incremental change (eq. 3)[1] (mg/kg)	Effect of deposition (eq. 2) (mg/kg)	Net effect (eq. 1f) (mg/kg)	Measured change (2005–10) (mg/kg)
				Minimal dispersion scenario					
PSNS015	17.7	2.7	39	1.533	0	no deposition	no deposition	0	+2.9
Steam plant	0.81	0.017	26	0.757	20.9	<0.001	-0.53	-0.53	-0.03
Pump 4	1.48	15	63	3.9	20.9	-1.74	-1.04	-2.77	-2.20
Pump 5	1.48	15	59	1.433	20.9	0.034	-0.34	-0.31	-1.21
Pump 6	2.31	16	43	0.89	20.9	1.09	-0.15	+0.93	-0.79
				Maximum dispersion scenario					
PSNS015	17.7	2.7				0.028			
Steam plant	0.81	0.017				<0.001			
Pump 4	1.48	15	OUB Marine	0.9	1,620	0.005	-0.67	-0.62	NA
Pump 5	1.48	15				0.005			
Pump 6	2.31	16				0.014			

[1] C_{wc}, concentration of total mercury of solids in OU B Marine water column = 0 23 mg/kg

The incremental change of STHg of each BNC discharge will generally be a positive value (indicating an increase in STHg concentration). However, the incremental change of STHg will be offset by the effect of deposition of suspended-solids expected from the sedimentation of present-day (2008–2009) solids suspended in the upper water column of OU B Marine (C_{wc}). C_{wc} was 0.23 mg/kg, or about 0.67 mg/kg lower than the average STHg concentration in OU B Marine before the USGS study (Paulson and others, 2012).

The calculated incremental change of STHg for each BNC source is shown in table 7, and ranges from a decrease in concentration in the cell receiving pump 4 solids (-1.74 mg/kg) to essentially zero change (PSNS015, steam plant, and pump 5) to an increase in concentration in the cell receiving pump 6 solids (+1.09 mg/kg). The negative estimated incremental change from pump 4 solids is because of the high STHg concentration ($C_{sediment}$ = 3.9 mg/kg) in cell 63.

Summing the incremental change with the effect of deposition resulted in a range of net effects. Net effects were estimated to be zero in the cell receiving PSNS015 solids because no deposition was assumed. However, the measured change from 2005–10 was +2.9 mg/kg, and the total organic carbon content also increased two-fold, suggesting that deposition of high THg solids may be occurring. Net effects were estimated to be negative (-0.53 mg/kg, indicating a decrease in STHg concentrations over time) in the cell receiving steam plant solids, while the measured change from 2005–10 was minimal (-0.03 mg/kg). In the cell receiving Dry Dock 1-5 solids discharged by pump 4, net effects were negative (-2.77 mg/kg), which compared well with the measured change from 2005–10 of -2.20 mg/kg. When solids from Dry Docks 1–5 were discharged to cell 59 from pump 5 rather than cell 63 from pump 4 (cell 63), net effects were slightly negative (-0.31 mg/kg), which was less negative than the measured change (-1.21 mg/kg). In the cell receiving Dry Dock 6 solids discharged by pump 6, net effects were estimated to be positive (+0.93 mg/kg), while the measured change was negative (-0.79 mg/kg).

Overall, the estimated changes in STHg concentrations were inconsistent with the sedimentation model under minimal dispersion scenario. There appears to be no correlation (Kendall tau = 0.33) between the net effects of high-THg solids and actual changes 2005–10 (table 7) for the minimal dispersion scenario. There are several physical and biogeochemical explanations why the application of the minimal dispersion scenario did not accurately predict the temporal changes in the sediment. The discharge pipes from the four sources are oriented horizontally, thus any momentum the discharge water conveys will tend to disperse the water and suspended solids over some distance from the discharge pumps. While the sump pumps provide the dry dock discharge water with considerable horizontal momentum at the submarine discharge location, the hydraulic head in PSNS015 can also impart horizontal momentum at its discharge point. The contrasting temporal scale of the discharge measurements and the monitoring of the sediment also provide difficulties in interpreting the results. The limited sampling of the dry dock discharges is expected to represent the present average annual conditions. However, the sediment for the monitoring program is collected to a depth of 10 cm, which may represent average sedimentation over the last 20–40 years (Paulson and others, 2010). In addition, the radiological data suggest that the sediment of Sinclair Inlet and the OU B Marine is completely mixed to a depth of 4 cm. If THg concentrations of suspended solids of discharge water have been decreasing because of pollution abatement programs initiated at the naval base, the full effect to decreasing solids in the discharge water may not be fully manifested for 10 years in the top 4 cm of sediment and up to 40 years in the top 10 cm used by the monitoring program. Sinclair Inlet is a physically dynamic system and sediment also may be resuspended and dispersed horizontally once it is initially deposited (Paulson and others, 2010). A number of physical factors may be affecting the fate of the solids discharged by these four sources. In particular, the relation between tidal flushing of stormwater drain PSNS015 and the dramatic increase of STHg in cell 39 needs further investigation.

Maximum Dispersion Scenario

Since the minimal dispersion scenario does not seem applicable to evaluating the effects of the four high-THg concentration-solids discharged from the BNC, the overall incremental change of STHg on OU B Marine sediment was evaluated by the maximum dispersion scenario. In the maximum dispersion scenario, each of the four high-THg concentrations are dispersed over the entire OU B Marine and mixed with 1,620 metric tons/yr of solids settling out of the water column. The 2.7 metric tons of solids containing 17.7 mg/kg THg flushed from PSNS015 would lead to the largest incremental change of +0.028 mg/kg in STHg concentrations in OU B Marine, while the incremental increase of the steam plant solids is estimated to be smallest (<0.001 mg/kg). The incremental changes of STHg for the dry dock discharges were estimated to be 0.005 and 0.014 mg/kg for Dry Docks 1–5 and 6, respectively. The sum of these incremental changes of STHg of these four sources was estimated to be +0.053 mg/kg of STHg throughout the entire area of OU B Marine. Thus, the discharge of the four high-THg concentrations solids would inhibit the recovery of STHg in OU B Marine sediments by settling of present day suspended solids (-0.67 mg/kg) by less than 10 percent.

Summary

The results of studies presented here and in Paulson and others (2012) indicate the most significant releases of both filtered total mercury (FTHg) and particulate total mercury are from soils in fill areas of Site 2 around the Puget Sound Naval Shipyard (PSNS)015 stormwater drain and Site 10C around well Operable Unit B Terrestrial (OUBT)-709. Another release of total mercury (THg) of solids from the Bremerton naval complex (BNC) of potential significance appears to be the relatively small loading of solids with high-THg concentrations discharged by the dry docks.

The tidal study at well OUBT-709 was conducted under the worst-case scenario of a king tide with maximum high higher tides for the year followed by minus lower low tides. The results indicate that the hydrologic connection of seawater with the contaminated soils of Site 10C occurs through the intertidal zone. Solids suspended in well OUBT-709, located about 30 meters inland, had some of the highest THg concentrations seen on the BNC. High FTHg concentrations in groundwater collected from this well were only observed for a short time after the elevation of Sinclair Inlet declined below the elevation of the water table at the well. If continued sampling of FTHg and THg of suspended solids in well OUBT-709 focused on the ebbing tidal cycle to lower low tide under a variety of tidal conditions, it may determine the prevalence of high FTHg concentrations at well OUBT-709. Intertidal studies, based on the temporal THg concentrations as a function of hydrologic gradient between the well and Sinclair Inlet, would provide data to quantify the annual aqueous- and particulate-phase THg released into Sinclair Inlet.

The source of the high-THg concentration solids discharged from the sumps of the dry docks is unknown. Of the four identified sources, suspended solids from ambient groundwater and suspended solids in Sinclair Inlet seawater seeping directly into the sumps have been eliminated as a possible source of these high-THg concentration solids. The solids suspended in groundwater flowing through contaminated soils, as measured at OUBT-709, is likely the most plausible source of these high-THg concentration solids. Measurement of THg concentrations of solids suspended in the process water discharged into the dry docks sump could eliminate process water as a possible source and allow environmental managers to focus on the groundwater transport to the sumps.

Acknowledgments

Current U.S. Geological Survey Washington Water Science Center (WAWSC) staff, James Foreman, and former staff, Cg Laird and Sarah Henneford, assisted in the sampling of the steam plant, dry docks and streams. Current U.S. Geological Survey WAWSC staff, Raegan Huffman, Steve Cox, Karen Payne, Patrick Moran, and Greg Justin, and former staff, Sarah Henneford and Rick Wagner, assisted in sampling groundwater. Wisconsin Mercury Research Laboratory (Middleton, Wisconsin) staff, David Krabbenhoft, Jake Ogorek, and Thomas Sabin analyzed the water and solids samples for THg. Former U.S. Geological Survey WAWSC staff, Terri Hurlbut, oversaw the processing of samples for TSS and SSC analyses, weighed filters, and calculated final concentrations.

Mark Wicklein of Naval Facilities Northwest and Jacquelyn Young of PSNS&IMF provided flow data for the industrial discharges. Duy Pham and Charles Parkins of PSNS&IMF and John Pittz of Naval Facilities Northwest provided logistical support for sampling within the BNC. Previous drafts of this report were reviewed by WAWSC staff Richard Dinicola, Mark Kozar, and Matt Bachmann, California Water Science Center staff, Jacob Fleck, and Dr. Robert Johnston of the U.S. Space and Naval Warfare Systems Command.

References Cited

Brandenberger, J.M., Louchouarn, P., Kuo, L-J., Crecelius, E.A., V. Cullinan, Gill, G.A., Garland, C., Williamson, J., and Dhammapala. R., 2010, Control of toxic chemicals in Puget Sound, phase 3–study of atmospheric deposition of air toxics to the surface of Puget Sound: Washington State Department of Ecology Report 10-02-012, Olympia, Wash, 90 p. and appendices.

ENVironmental inVESTment, 2006, Puget Sound Naval Shipyard & Intermediate Maintenance Facility project ENVVEST community update June 2006 (Brochure and CD): Marine Environmental Support Office-NW, Space and Naval Warfare Systems Center, Bremerton, Wash., August 2006, Ecology Publication Number 06-10-54, accessed on January 7, 2013 at http://environ.spawar.navy.mil/Projects/ENVVEST/ENVVEST2006/index.html.

Huffman, R.L., Wagner, R.J., Toft, J., Cordell, J., DeWild, J.F., Dinicola, R.S., Aiken, G.R., Krabbenhoft, D.P., Marvin-DiPasquale, M., Steward, A.R., Moran, P.W., and Paulson, A.J., 2012, Mercury species and other selected constituent concentrations in water, sediment, and biota of Sinclair Inlet, Kitsap County, Washington, 2007–10: U.S. Geological Survey Data Series 658, 64 p.

Lewis, M.E., and Brigham, M.E., 2004, Low-Level mercury, National Field Manual for the collection of water-quality data: U.S. Geological Survey Techniques of Water-Resources Investigations, book 9, sec. 6.4B, chap. A5, 26 p.

Malins, D.C., McCain, B.B., Brown, D.W., Sparks, A.K, Hodgins, H.,1982, Chemical contaminants and abnormalities in fish and invertebrates from Puget Sound: National Oceanic and Atmosperic Administration Technical Memorandum OMPA-19. Boulder, Colo., 168 p.

Paulson, A.J., Keys, M.E., and Scholting, K.L., 2010, Mercury in sediment, water, and biota of Sinclair Inlet, Puget Sound, Washington, 1989–2007: U.S. Geological Survey Open-File Report 2009-1285, 220 p.

Paulson, A.J., Dinicola, R.S., Noble, M.A., Wagner, R.J., Huffman, R.L., Moran, P.W., and DeWild, J.F., 2012, Sources and sinks of filtered total mercury and concentrations of total mercury of solids and of filtered methylmercury, Sinclair Inlet, Kitsap County, Washington, 2007–10: U.S. Geological Survey Scientific Investigations Report 2012-5223, 94 p.

Prych, E.A., 1997, Numerical smulation of ground-water flow paths and discharge locations at the Puget Sound Naval Shipyard, Bremerton, Washington. U.S. Geological Survey, Water-Resource Investigations Report 96-4147, 43 p.

Skahill, B.E., and LaHatte, C., 2007, Hydrological simulation program–FORTRAN modeling of the Sinclair–Dyes Inlet watershed for the Puget Sound Naval Shipyard & Intermediate Maintenance Facility Environmental Investment Project - FY 2007 Report: U.S. Army Engineer Research and Development Center, Waterways Experiment Station, Vicksburg, Miss., Report to the U.S. Navy Puget Sound Naval Shipyard and Intermediate Maintenance Facility Environmental Division, 59 p. (Also available at: http://environ.spawar navy mil/Projects/ENVVEST/FC Model_Report/HSPF_reports htm.)

U.S. Environmental Protection Agency, 2000, EPA superfund record of decision–Puget Sound Naval Shipyard Complex: Seattle, Wash., Environmental Protection Agency ID: WA2170023418 OU 02 Bremerton, Wash., EPA/ROD/R10-00/516, Region X.

U.S. Navy, 1992, Site Inspection Report: Final report prepared by the URS team under contract #N62474-89-0-9295, v. 1, 2,063 p.

U.S. Navy, 2002, Final remedial investigation report operable unit B, Bremerton naval complex, Bremerton, Washington: Final report prepared by the URS Grier under contract #N62474-89-0-9295, v. 1, 1,402 p.

U.S. Navy, 2006a, 2003 marine monitoring report OU B marine, Bremerton naval complex Bremerton, Washington: Final report prepared by the URS Group, Inc. under contract No. N44255-02-D-2008, 295 p.

U.S. Navy, 2006b, 2005 marine monitoring report OU B marine, Bremerton naval complex Bremerton, Washington: Final Report prepared by the URS Group., Inc. under contract No. N44255-02-D-2008, 272 p.

U.S. Navy, 2007b, Sediment transport study and natural recovery model report, Bremerton naval complex, Bremerton, Washington: Seattle, Wash., URS Group., Inc., 224 p.

U.S. Navy, 2008a, Second five-year review, Bremerton naval complex, Bremerton, Washington: Silverdale, Wash., U.S. Navy Naval Facilities Engineering Command Northwest, 258 p.

U.S. Navy, 2008b, 2007 marine monitoring report OU B marine, Bremerton naval complex, Bremerton, Washington: Final report prepared by the URS Group, Inc. under contract No. N44255-05-D-5100.

U.S. Navy, 2011, 2010 Preliminary trend analyses, Bremerton naval complex: Bremerton, Washington, Draft 2010 Marine Monitoring Report, appendix B.

U.S. Navy, 2012, All known, available, and reasonable methods of treatment study—Puget Sound Naval Shipyard & Intermediate Maintenance Facility Bremerton, Washington: Naval Facilities Engineering Command Northwest, Silverdale, Washington, and Puget Sound Naval Shipyard & Intermediate Maintenance Facility, Bremerton, Washington, 278 p.

Appendix A. Sampling Details and Quality Assurance Results

Table A1. Quality assurance data for filtered, particulate, and unfiltered total mercury, Sinclair Inlet, Kitsap County, Washington.

[**Abbreviations:** Conc., concentration; mg/L, milligrams per liter; RPD, relative percent difference; ng/L, nanograms per liter; WAWSC, Washington Water Science Center; OUBT, Operable Unit B-Terrestrial, GFF, glass fiber filter; %, percent; –, not analyzed]

Sample date	Sample time	Device or site	Total suspended solids		Particulate total mercury		Filtered total mercury		Total mercury in unfiltered water	Comments on procedure
			Conc. (mg/L)	RPD (%)	Conc. (ng/L)	RPD (%)	Conc. (ng/L)	RPD (%)	Conc. (ng/L)	
						Source water				
10-17-11	1518	WAWSC Purelab® water	–	–	–	–	–	–	0.08	–
05-10-12	1235	WAWSC Purelab® water	–	–	–	–	–	–	<0.04	After changing cartridges and filters
					Pre-testing Teflon churn splitter water					
10-17-11	1408	"Spokane" churn	–	–	<.019	–	0.22		–	–
10-17-11	1418	"Tacoma A" churn	–	–	<.024	–	0.20		–	–
10-17-11	1428	"Tacoma B" churn	–	–	<.040	–	0.04		–	–
10-25-11	1008	"Tacoma C" churn	–	–	<.035	–	–		–	–
					Pre-testing polypropylene churn splitter water					
11-15-11	1508	"Spokane 3" churn	–	–	<.034	–	–	–	–	–
11-29-11	1108	"Spokane 5" churn	–	–	<.023	–	–	–	–	–
11-29-11	1208	"Tacoma 11" churn	–	–	<.025	–	–	–	–	–
11-29-11	1308	"Tacoma 2" churn	–	–	<.030	–	–	–	–	–
11-29-11	1418	"1" churn	–	–	<.025	–	0.06	–	0.85	Through spigot
11-29-11	1408	"1" churn	–	–	<.025	–	–	–	1.2	Not through spigot
11-29-11	1508	"Spokane 2" churn	–	–	–	–	–	–	0.2	Not through spigot
05-10-12	1240	"Spokane 4" churn	–	–	–	–	–	–	0.38	Through all-plastic spigot
						Field blanks				
11-27-11	0715	OUBT-709	–	–	–	–	0.09	–	–	Through sampling port, tubing, Meissner field-filtered
12-26-11	1547	OUBT-724	–	–	–	–	0.11	–	–	Through sampling port, tubing, Meissner field-filtered
02-29-11	1201	Annapolis Creek	[1]0.12	8.1	<.072	–	0.08	–	–	Through tubing, field bottle, GFF lab-filtered
02-29-11	1202	Annapolis Creek	[1]0.13		–	–	–	–	–	Through tubing, field bottle, GFF lab-filtered
05-07-12	0910	BNC stormwater blank	[1]0.07	–	–	–	–	–	–	Through port, tubing, field bottle, GFF lab-filtered

Table A1. Quality assurance data for filtered, particulate, and unfiltered total mercury, Sinclair Inlet, Kitsap County, Washington.—Continued

[**Abbreviations:** Conc., concentration; mg/L, milligrams per liter; RPD, relative percent difference; ng/L, nanograms per liter; WAWSC, Washington Water Science Center; OUBT, Operable Unit B-Terrestrial, GFF, glass fiber filter; %, percent; –, not analyzed]

Sample date	Sample time	Device or site	Total suspended solids		Particulate total mercury		Filtered total mercury		Total mercury in unfiltered water	Comments on procedure
			Conc. (mg/L)	RPD (%)	Conc. (ng/L)	RPD (%)	Conc. (ng/L)	RPD (%)	Conc. (ng/L)	
Cross-over blanks										
11-27-11	1520	OUBT-709, water level	–	–	–	–	5.33	–	–	Through sampling port, tubing, Meissner field-filtered
12-13-11	1711	Field-rinsed used Teflon churn	[1]0.13	–	0.03	–	–	–	–	–
12-26-11	1547	OUBT-724, water level	–	–	–	–	0.73	–	–	Through sampling port, tubing, Meissner field-filtered
Laboratory replicates										
11-27-11	1105	Vault (O/43-9) of	8.42	8.0	–	–	–	–	–	–
	1106	PSNS82.5	9.12		–	–	–	–	–	–
11-27-11	1148	Catch basin 3231	8.33	39	–	–	–	–	–	–
	1149	of PSNS82.4	12.41		–	–	–	–	–	–
11-27-11	0119	OUBT-709 water	22.70	6.2	–	–	–	–	–	–
	0120	level	24.16		–	–	–	–	–	–
Field replicates										
11-28-11	0035	OUBT-709, water	–	–	–	–	70.0	4.5	–	–
	0035	level	–	–	–	–	73.2		–	–
12-26-011	2102	OUBT-724, water	–	–	–	–	4.49	1.5	–	–
	2103	level	–	–	–	–	4.56		–	–
12-26-11	2258	OUBT-724, water	–	–	–	–	4.04	4.4	–	–
	2259	level	–	–	–	–	4.22		–	–
02-29-12	1200	Gorst Creek	9.53	1.7	–	–	–	–	–	–
	1201		9.38		–	–	–	–	–	–
02-29-12	1345	Anderson Creek	23.18	8.6	–	–	–	–	–	–
	1346		21.26		–	–	–	–	–	–
02-29-12	1500	Anderson Creek	13.52	0.4	–	–	–	–	–	–
	1501		13.47		–	–	–	–	–	–
02-29-12	1645	Blackjack Creek	28.59	2.0	–	–	–	–	–	–
	1646		29.17		–	–	–	–	–	–
03-14-12	1115	Anderson Creek	17.10	4.7	2.02	6.6	1.77	10.7	–	–
	1116		16.33		1.89		1.97		–	–
Sampling method comparisons										
11-27-11	0754	OUBT-709, bottom	–	–	–	–	31.7	24	–	Meissner field-filtered
	0757	sampling port	–	–	–	–	40.4		–	Collected in bottle and GFF lab-filtered
12-13-11	1001	Steam plant effluent	–	–	0.104	43	–	–	–	Collected in bottle and GFF lab-filtered
	1011		–	–	0.067		–	–	–	Collected in bottle from churn splitter, GFF lab-filtered

[1] Based on a nominal filtered volume of 1 liter for low total suspended solids concentration samples.

Table A2.   Loading calculation of filtered total mercury using wet and dry season data, Sinclair Inlet, Kitsap County, Washington.

[**Abbreviations:** ng/L, nanograms per liter; m³/s, cubic meters per second; m³, cubic meters; g/yr, grams per year; –, not available]

Season and area	Average filtered total mercury concentration (ng/L)	Average discharge (m³/s)	Total annual volume (million m³)	Filterered total mercury loading (g/yr)
Wet season	–	2.104	38.8	72.5
Total gaged area	–	1.988	36.5	67.3
Gorst Creek	1.11	0.93	17.1	19.0
Anderson Creek	2.05	0.32	5.9	12.1
Blackjack Creek	2.67	0.72	13.2	35.2
Annapolis Creek	3.33	0.018	0.3	1.0
Ungaged areas	2.28	0.116	2.3	5.2
Dry season	0.57	0.75	9.9	5.6
Annual	–	–	48.7	78.1

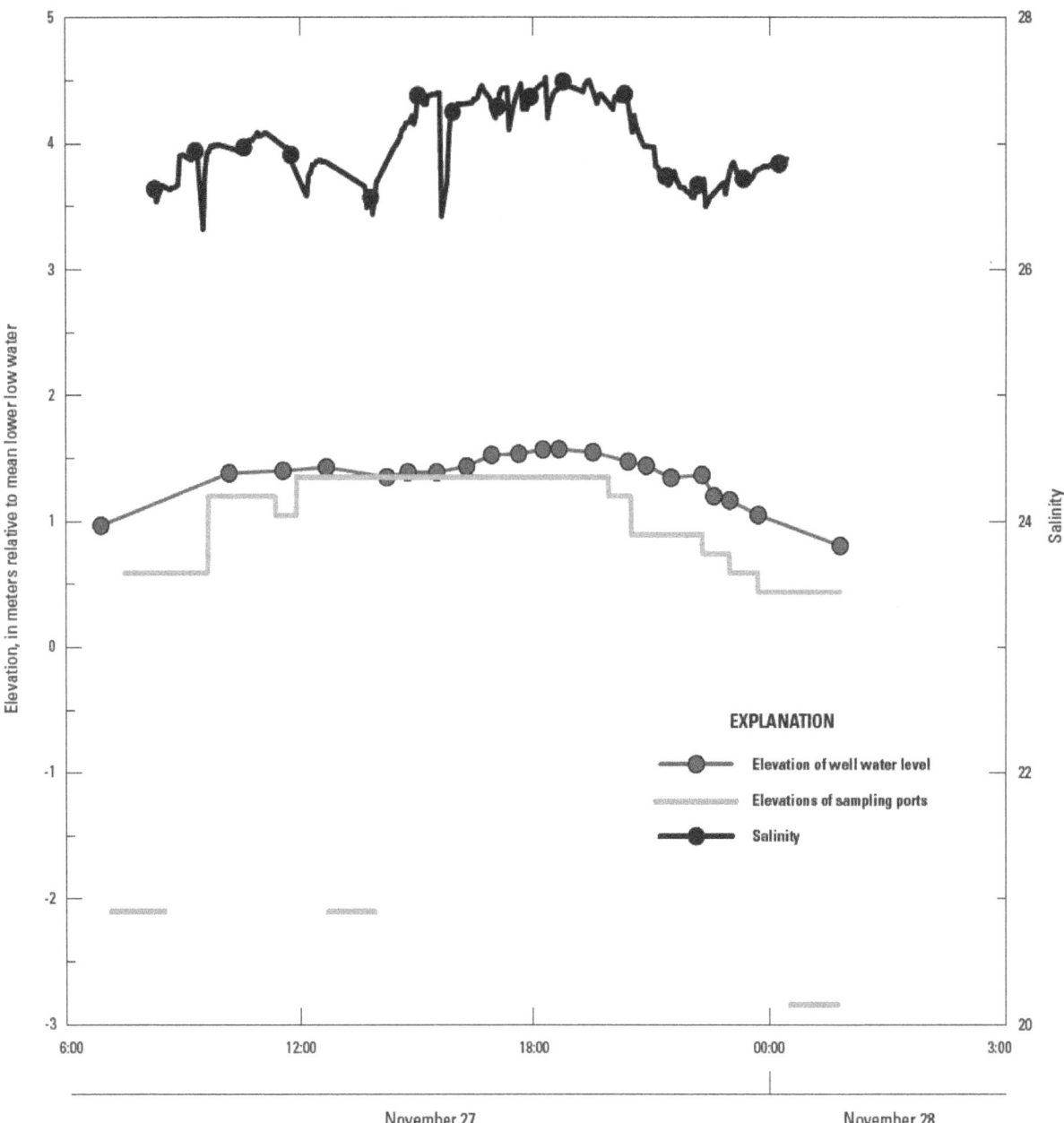

Figure A1.   Elevations of well OUBT-709 water levels and sampling ports at the Bremerton naval complex, Sinclair Inlet, Kitsap County, Washington, November 27–28, 2011. Also shown are salinity data from the sampling port just below the water table.

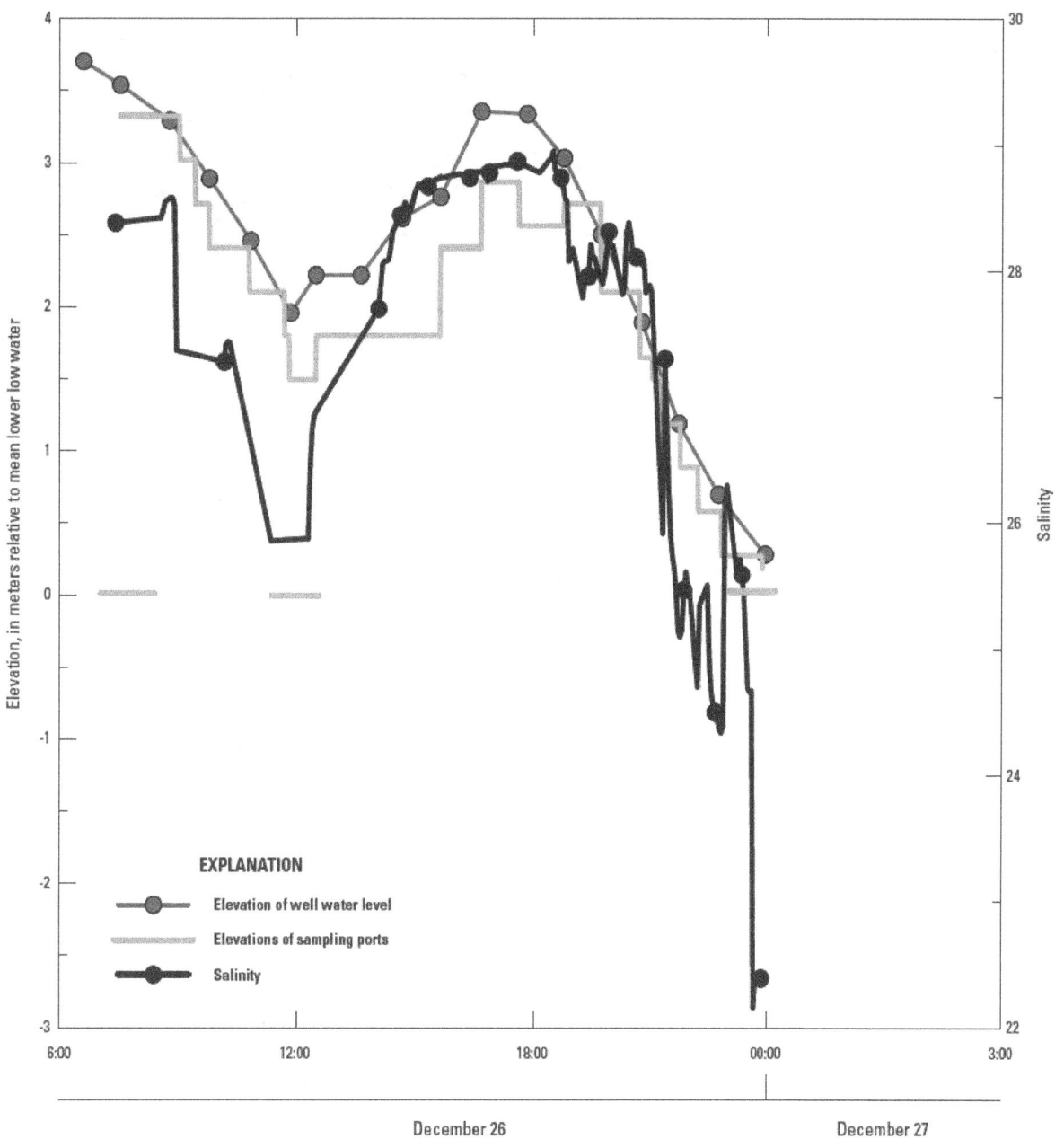

Figure A2.   Elevations of well OUBT-724 water levels and sampling ports at the Bremerton naval complex, Sinclair Inlet, Kitsap County, Washington, December 26–27, 2011. Also shown are salinity data the sampling port just below the water table.

www.ingramcontent.com/pod-product-compliance
Lightning Source LLC
Chambersburg PA
CBHW081405170526
45166CB00010B/3218